**Your Best Options for
Solar Heating and Cooling,
Wood, Wind, and Photovoltaics**

Home
Energy

by Dan Halacy

Edited by Carol Hupping

Design: Barbara Field
Illustration: Frank Fretz, Brian Swisher
Editorial Assistance: Cheryl Winters Tetreau,
            Margaret J. Balitas,
            Bobbie Hartranft, Joe Carter

Rodale Press, Emmaus, Pa.

Home Energy

Printed in the United States of America on recycled paper containing a high percentage of de-inked fiber.

Cover photographs clockwise from left to right by Carl Doney, John P. Hamel, Michael Kanouff, Mitch Mandel, and Carl Doney

Library of Congress Cataloging in Publication Data

Halacy, D. S. (Daniel Stephen), 1919–
    Home energy.

    Includes index.
    1. Dwellings—Heating and ventilation. 2. Dwellings—
Air conditioning. 3. Renewable energy sources.
I. Hupping, Carol. II. Title.
TH7226.H35 1984   644'.1   83-24778
ISBN 0-87857-493-X   hardcover

2 4 6 8 10 9 7 5 3 1 hardcover

Contents

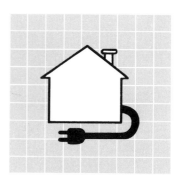
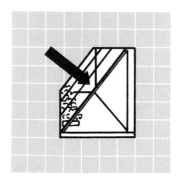

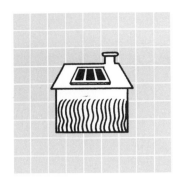

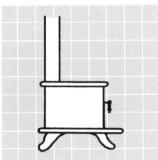

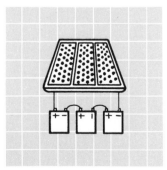

Introduction

The beginnings of *Home Energy* go back more than 10 years, when the book that inspired it was published. That book, *Producing Your Own Power,* marked the start of what was to become known as the "alternative energy boom." For it was then, in the early '70s, that people began to see that there were legitimate modern home energy and heating sources other than coal, oil, gas, and nuclear power.

Producing Your Own Power was the first book about alternative home energy, but it was hardly the last. Many books followed and so, too, did more designs and hardware. The alternative energy industry blossomed; wood stoves and solar collectors became big business. Millions of homeowners bought into this "new" technology, some for the novelty of it, and others for the anticipated energy savings.

Luckily, as in all new fields, most of the poor products couldn't stand the competition, but the best of them made it through. Innovations and refinements ultimately led to improvements and much better know-how. For instance, we know that:

Solar water heaters are much more reliable, efficient, and durable now. And we've learned that they don't have to face exactly due south to work well, making installation attractive and convenient

Passive solar heating has proved even more effective than early designers thought it could be. Good passive options can typically save 50 percent and in some cases almost 100 percent of heating costs

Airtight wood stoves, originally imported from Scandinavia in the mid-'70s, revolutionized wood heating and have made traditional stoves just about obsolete. Fireplaces, formerly energy losers, can now be retrofitted with well-designed inserts that make them attractive, efficient heating devices. The new catalytic combustors for wood stoves and fireplaces minimize airborne pollutants and also boost efficiency

Out of the creative aerodynamic engineering of the '70s have come wind generator designs that perform efficiently where wind blows strong and steadily. They are most cost-effective in remote areas where bringing in power lines is prohibitively expensive. But the new Public Utility Regulatory Policies Act (PURPA), which requires utility companies to buy power from homes that generate excess electricity, may make wind-utility-connected systems practical even in some suburban areas

Solar electricity produced by solar cells, or photovoltaic (PV) cells, is the most exciting renewable energy development of the '80s and will prove to be even more exciting as the century comes to a close. Affordable only in spacecraft a decade ago, PV can now be cost-effective in remote areas. Utility-connected systems, thanks to PURPA, are already working for some homeowners as a convenient backup power supply

In 1974, when *Producing Your Own Power* was published, the enthusiasm for alternative energy was there, but we had a lot to learn. These last 10 years we *have* learned a lot, and this new book is a product of that progress.

Home Energy is not a rehash of *Producing Your Own Power,* nor of any of the other, earlier books. As you no doubt knew before you picked up this book, or at least have discovered in the last few paragraphs, things have been changing. You'll find in *Home Energy* only those designs and systems and applications that make sense now, here in the mid-'80s. *Options* is a key word in this book, because *Home Energy*'s job is to present you with the best of today's options in solar, wood, and wind—in terms of cost-effectiveness, performance, convenience, reliability, and attractiveness. Once you have an idea of what might be best for you, this book will give you enough information to narrow down your choices, based on your budget, your location, and your expectations.

Home Energy will help you sort through the abundance of information on energy conservation and energy-efficient appliances and heating and cooling systems, and on everything solar, all types of wood stoves and fireplaces and their assorted apparatus. It's meant to explain the potential and limitations of wind generating systems, and introduce you to what very well might prove to be the most useful alternative energy source of them all: photovoltaics. The chapters are arranged in general order of their usefulness, simplicity of application, and cost-effectiveness.

Before you start thinking about supplementing or replacing your present energy source, you've got to think about cutting down your use of electricity and reducing your heating and cooling needs. That's why Chapter 1 is devoted to energy conservation. Shape up what you've got, then build upon that. If you use nothing more than this first chapter, you'll get your money's worth out of this book, and then some.

Subsequent chapters each take you through a specific alternative energy and its applications. Chapter 2 tells you about solar water heating with active systems like rooftop collectors and pumps to move water to storage tanks and your faucets. It also explains how to use passive systems, like breadbox collectors and thermosiphon systems, to heat water. In Chapter 3 you'll find out about how to heat and cool your house with passive solar energy systems, like sunspaces, Trombe walls, and window box collectors. Here you'll also learn about natural cooling designs like thermal chimneys and cool tubes. Chapter 4 focuses on active solar space heating, similar to active solar water heating in that it uses fans and pumps to move hot air (or hot water, which in turn heats the air) from collectors to living spaces.

The options for wind generating systems for both stand-alone setups that rely on battery storage and for utility-connected home systems that make use of PURPA can be found in Chapter 5. Chapter 6 is devoted to wood stoves, central wood heating, and wood-heated water. And the last chapter, Chapter 7, explains home applications of photovoltaics that can change sunlight (or any light for that matter) directly into electricity.

Some Tips for Getting the Most from This Book

Home Energy has been designed to make your energy decisions as easy as possible. You'll find guides and shortcuts to information throughout.

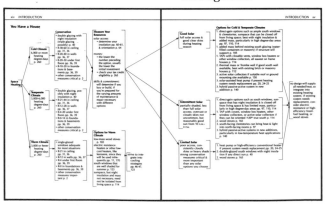

The first of these are decision trees: these are special charts that take you down a choice of paths, or along "branches," depending upon your resources (i.e., access to good sun, wind, plentiful wood) and your needs (i.e., supplemental electricity, space heating, hot water, cooling), and then lead you to viable options that are page-referenced to specific sections of the book that describe those options in greater detail. These can be found on the next few pages.

Another tool for easy information access can be found at the start of each chapter. Here you'll find a list of chapter highlights. If you want a very quick and very brief look at some of the major points in the chapter, you can read this list. If something catches your eye, you can note the page reference immediately following the highlight and turn to that page for more reading.

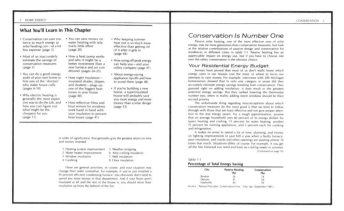

For a longer review of the book, skim the art captions. They have been designed to emphasize and elaborate upon the text. They illustrate:

The major components of a system

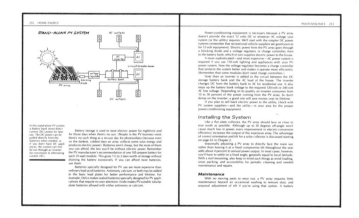

How a piece of equipment or a system works

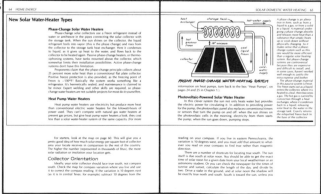

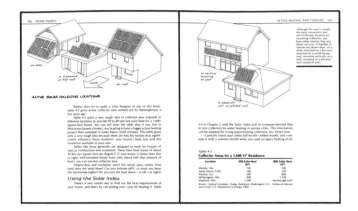

Variations on a theme

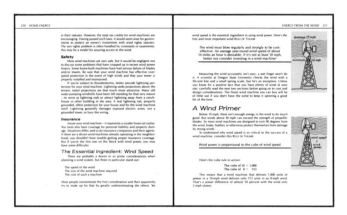

RULES OF THUMB are placed throughout the book. Most of them provide you with a short and simple design or consumer guideline, such as how to size a water-storage tank, how much south-facing glass your solar addition needs, how big a wood stove you should buy.

And to wrap it all up, at the end of each chapter there is a list of suggested books to read, with some comments about what you'll find helpful in each. Once you've narrowed down your energy options, these books will give you more specific, more detailed facts about these options.

The Decision Trees

Flip the page and you'll find decision trees that will help you zero in on those energy options that will probably be of most interest to you. No tool like this, which is designed to meet some basic needs of *everyone*, can really meet the personal, very specific needs of *anyone*, but what the decision trees can do for you is lead you step by step through the first things anyone contemplating a home energy invest-ment should think about before plunging ahead.

Glance through this book, if you haven't done so already. You'll be struck by the great number of energy options to be found here, options that, on face value, are open to you. If you're new to solar and wind and wood, these choices might seem a bit overwhelming. But the more you know about them, and the more you can pin down your own needs, the less confusing these options become. What you'll soon find is that some aren't really options for you at all, and among the ones that are left, some will certainly be better for you than others.

These decision trees are here to take you through this general selection process before you get into the book, so that you don't waste a lot of time on options that probably won't be good for you in the long run. Once you've reached the end of the "branches" and have gotten to your most probable best choices, you'll find numbers that correspond to the pages later on in the book that give you more information about those specific options.

If you have a house already, your situation, naturally, is going to be different from what it would be if you were planning a new house, so there are four trees that work with existing houses, and another four trees for new houses. If you want to improve the energy picture for your present house, look at the first eight pages. These deal with heating and cooling choices, heating your water, and providing elec-tricity. The second set of eight pages is for you if you're contemplating a new house.

Begin using the trees by deciding what your greatest needs are: an alternative or supplemental way to heat your house, and/or a more efficient way to cool it? Perhaps water heating is your priority. Or a way to supplement your utility-provided electricity. Take your pick and move from left to right along the branches of the appropriate tree. Your first stop along all the branches of the trees will be conserva-tion, for no matter what your needs, eliminating wasted or unnecessary energy will make your final options more successful.

You may find that your climate, natural resources (like available solar radiation, ready firewood, or wind speed in your backyard), and budget afford you the luxury of using many of the options at the end of the branches. But more than likely, your particular situation will lead you to one or only a few practical options. You may also find that you have several choices, but that most (or maybe even all, especially if you already have the house) will work most effectively as supplements to conventional home energy systems.

You Have a House

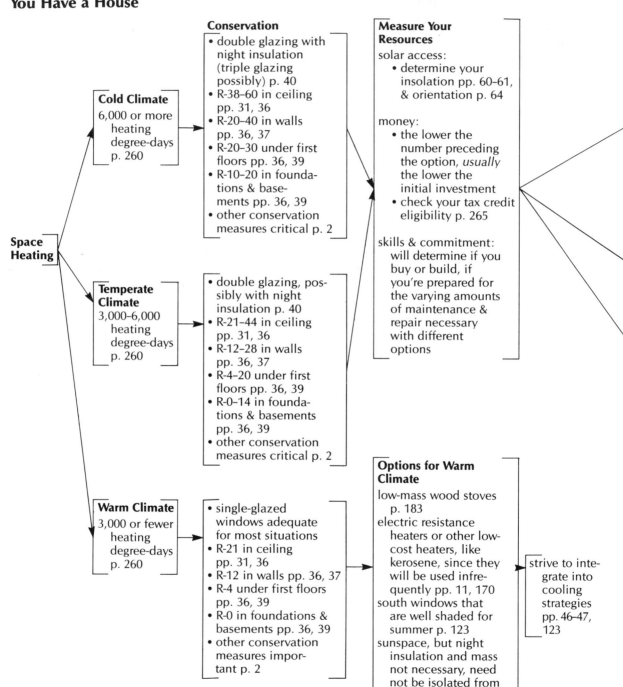

Space Heating

Cold Climate
6,000 or more heating degree-days p. 260

Temperate Climate
3,000–6,000 heating degree-days p. 260

Warm Climate
3,000 or fewer heating degree-days p. 260

Conservation
- double glazing with night insulation (triple glazing possibly) p. 40
- R-38–60 in ceiling pp. 31, 36
- R-20–40 in walls pp. 36, 37
- R-20–30 under first floors pp. 36, 39
- R-10–20 in foundations & basements pp. 36, 39
- other conservation measures critical p. 2

- double glazing, possibly with night insulation p. 40
- R-21–44 in ceiling pp. 31, 36
- R-12–28 in walls pp. 36, 37
- R-4–20 under first floors pp. 36, 39
- R-0–14 in foundations & basements pp. 36, 39
- other conservation measures critical p. 2

- single-glazed windows adequate for most situations
- R-21 in ceiling pp. 31, 36
- R-12 in walls pp. 36, 37
- R-4 under first floors pp. 36, 39
- R-0 in foundations & basements pp. 36, 39
- other conservation measures important p. 2

Measure Your Resources

solar access:
- determine your insolation pp. 60–61, & orientation p. 64

money:
- the lower the number preceding the option, *usually* the lower the initial investment
- check your tax credit eligibility p. 265

skills & commitment: will determine if you buy or build, if you're prepared for the varying amounts of maintenance & repair necessary with different options

Options for Warm Climate

low-mass wood stoves p. 183
electric resistance heaters or other low-cost heaters, like kerosene, since they will be used infrequently pp. 11, 170
south windows that are well shaded for summer p. 123
sunspace, but night insulation and mass not necessary, need not be isolated from living space p. 114

strive to integrate into cooling strategies pp. 46–47, 123

Good Solar

full solar access &
good clear skies
during heating
season

Options for Cold & Temperate Climates

1. direct-gain options such as ample south windows
   & clerestories; sunspace that can be closed off
   from living space; best with night insulation &
   added mass, particularly in high degree-day areas
   pp. 97, 110, 114
1. added mass behind existing south glazing (water-
   filled containers or masonry) if structure will
   support p. 105
1. TAPs with closable vents, window box heaters or
   other window collectors, all easiest on frame
   houses p. 114
2. thermal mass or Trombe wall if good south wall
   available, best with existing brick or masonry
   houses p. 112
3. active solar collectors if suitable roof or ground
   mounting site available p. 150
3. solar-assisted heat pump if present heating
   system needs replacement pp. 20, 24–25
3. hybrid passive-active system in new
   addition p. 140

Intermittent Solar

partially shaded, less
than full solar
access; overcast or
cloudy skies not
uncommon, but
reasonably good
sun from 10 A.M.–
4 P.M.

1. direct-gain options such as south windows; sunspace that has night insulation & is closed off
   from living space & has limited mass, particularly in high degree-day areas pp. 97, 110, 114
1. TAP with covers, window box heaters, other
   window collectors, or active solar collectors if
   they can be oriented ±20° true south p. 114
2. wood stoves p. 165
3. south-facing clerestories can bring heat & light
   into north-facing rooms p. 97
3. hybrid passive-active systems in new additions,
   particularly in low-temperature heat applications
   p. 140

Limited Solar

poor access, con-
sistently cloudy
skies or heavy shade
strong conservation
measures critical &
more important
than any solar
options you choose

1. heat pump or high-efficiency conventional heater
   if present system needs replacement pp. 20, 24–25
2. double-glazed south windows with night insulation if any direct sun p. 40
2. wood stoves p. 165

no design will supply
all needed heat, so
integrate into
existing heating
system; if existing
system needs
replacement, con-
sider electric
resistance or high-
efficiency fossil-
fuel heating, or
wood stoves

You Have a House

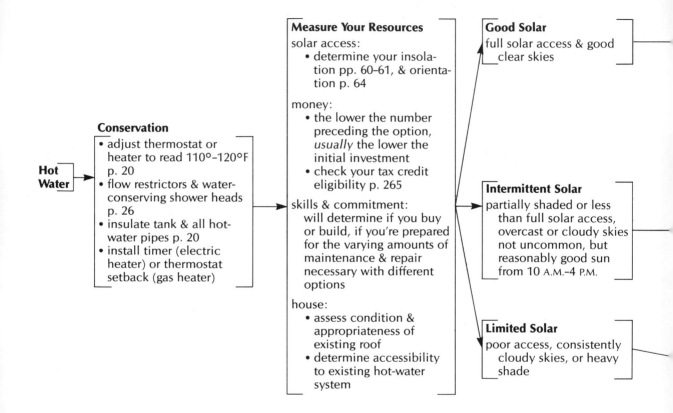

Hot Water

Conservation
- adjust thermostat or heater to read 110°–120°F p. 20
- flow restrictors & water-conserving shower heads p. 26
- insulate tank & all hot-water pipes p. 20
- install timer (electric heater) or thermostat setback (gas heater)

Measure Your Resources
solar access:
- determine your insolation pp. 60–61, & orientation p. 64

money:
- the lower the number preceding the option, *usually* the lower the initial investment
- check your tax credit eligibility p. 265

skills & commitment:
will determine if you buy or build, if you're prepared for the varying amounts of maintenance & repair necessary with different options

house:
- assess condition & appropriateness of existing roof
- determine accessibility to existing hot-water system

Good Solar
full solar access & good clear skies

Intermittent Solar
partially shaded or less than full solar access, overcast or cloudy skies not uncommon, but reasonably good sun from 10 A.M.–4 P.M.

Limited Solar
poor access, consistently cloudy skies, or heavy shade

Options

1. batch heater; easiest owner-built option;
   good roof support critical because of weight,
   otherwise use ground mount pp. 62, 81
2. thermosiphoning heater with freeze
   protection pp. 62, 76
3. active heater with freeze protection pp. 62, 76
2. solar-assisted heat pump for hot water &
   space heating if present heating & hot-water
   system needs replacement pp. 24–25
   pool and/or hot tub solar heating is probably
   possible in warm & temperate climates; pool
   cover recommended p. 82

any as preheater/booster for
existing water heating p. 75

1. batch heater; easiest owner-built option;
   good roof support critical because of
   weight, otherwise use ground mount;
   consider insulating cover pp. 62, 81
2. thermosiphoning heater with freeze
   protection; consider insulating cover pp. 62, 76
3. active heater with freeze protection pp. 62, 76
2. solar-assisted or conventional heat pump
   for hot water & space heating if present
   heating & hot-water system needs replacement pp. 24–25
1. high-efficiency, well-insulated conventional
   system if present one needs replacement
   p. 20

preheater/booster for existing
water heating p. 75

whatever design you choose,
pick a system that doesn't
depend upon high tem-
peratures to operate
efficiently

1. conservation is your best bet
1. upgrade water heater by insulating it well
   or replace with high-efficiency unit p. 20
1. heat pump for hot water & space heating if
   present heating & hot-water system needs
   replacement pp. 24–25

You Have a House

Cooling →

Conservation
- good attic ventilation (fan and/or roof vent) p. 44
- good cross-ventilation with operable windows on facing exterior walls p. 127
- window overhangs, awnings, trellises, shades p. 123
- good insulation and weather stripping as for space heating pp. 46–47
- vent and shade existing skylights, sunspaces pp. 46–47
- plant deciduous shade trees; vegetation rather than asphalt, stone, or concrete around house p. 123
- minimize exposed south windows in summer with reflective film or reflective insulation panels, especially in warm climates p. 41
- paint house and roof a light color, especially in warm climates p. 123

→

Measure Your Resources

money:
    the lower the number preceding the options, *usually* the lower the initial investment

skills & commitment:
    both will determine if you buy or build, if you're prepared for the varying amounts of maintenance & repair necessary with different options

Warm Climate
cooling degree-days
1,500–4,500 p. 260

Options
1. energy-efficient cooling equipment when replacing air conditioner p. 160
1. heat pump for heating & cooling, when replacing existing heating & cooling systems pp. 24–25
1. whole-house fan or overhead vents to bring in cool air from basement or shaded north windows or porch p. 44
3. if dry climate, evaporative & swamp coolers could be considered in place of conventional air conditioner p. 45
3. extra mass—such as extra masonry in walls or floors, or water-filled tanks—can absorb heat and prevent sharp temperature rises during the day p. 127

Temperate Climate
cooling degree-days
500–1,500 p. 260

1. small room fans with good cross-ventilation p. 127
2. whole-house fan to draw air up from basement or shaded north windows or porch p. 44

You Have a House

Electricity

Conservation
- minimize electric usage for space heating & hot water whenever possible p. 44
- see cooling for alternatives to air conditioning pp. 46–47
- install energy-efficient appliances & light fixtures p. 49; turn off when not in use
- if off-peak rates are available from your utility, install timers to automatically run dishwashers & clothes dryers during off-peak times p. 47

Measure Your Resources

money:
- be aware that any wind or PV options will usually involve a greater initial investment than utility power will
- check your tax credit eligibility p. 265

skills & commitment:
- both will determine if you buy or build, if you're prepared for the varying amounts of maintenance & repair necessary with different options

solar:
- determine your insolation pp. 60–61, & orientation p. 64

wind:
- determine wind speed p. 212
- check for obstructions p. 216–17
- check possible zoning restrictions p. 209

Remote Site (running in power lines is prohibitively expensive)

- available electricity will be limited, so conservation is critical; limit needs by using other power sources whenever possible
- don't use electricity for space or water heating, refrigerator, air conditioning, or other major uses
- not cost-competitive with fuels like wood & propane heating & oil lamps, although getting such fuels to remote site may be difficult

Good Sun
full solar access, good clear skies

Options

photovoltaics stand-alone system with enough batteries for a few days' storage p. 246

photovoltaics/gen-set hybrid system p. 248

your good solar might be better used for space heating & hot water pp. 57, 91

Good Wind
steady 10-mph winds with no major obstructions or zoning restrictions

wind power stand-alone system with enough batteries for a few days' storage p. 220

Utility hookup exists

- utility power today is your cheaper, more reliable electricity source; stick with it unless using renewable energy is more important than first cost & maintenance considerations

Good Sun
full solar access, good clear skies

photovoltaics with utility interconnection, if system can meet utility regulations p. 245

your good solar might be better used for space heating & hot water pp. 57, 91

Good Wind
steady 10-mph winds with no major obstructions or zoning restrictions

wind power with utility interconnection, if system can meet utility regulations p. 216

You Are Buying/Building a New House

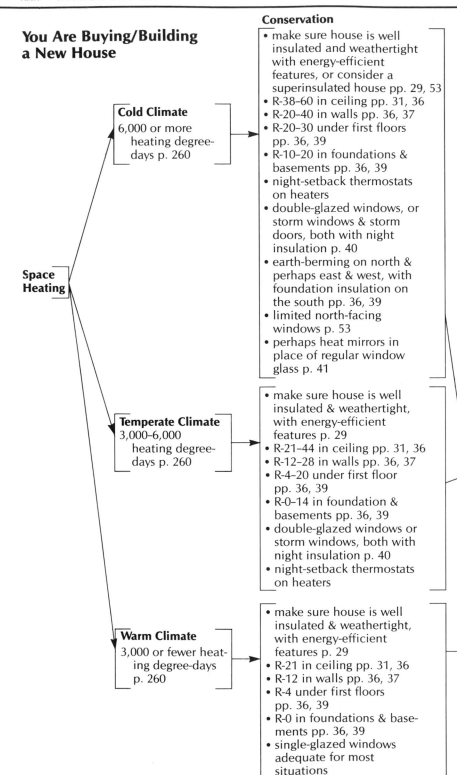

Space Heating

Cold Climate
6,000 or more heating degree-days p. 260

Temperate Climate
3,000–6,000 heating degree-days p. 260

Warm Climate
3,000 or fewer heating degree-days p. 260

Conservation

- make sure house is well insulated and weathertight with energy-efficient features, or consider a superinsulated house pp. 29, 53
- R-38–60 in ceiling pp. 31, 36
- R-20–40 in walls pp. 36, 37
- R-20–30 under first floors pp. 36, 39
- R-10–20 in foundations & basements pp. 36, 39
- night-setback thermostats on heaters
- double-glazed windows, or storm windows & storm doors, both with night insulation p. 40
- earth-berming on north & perhaps east & west, with foundation insulation on the south pp. 36, 39
- limited north-facing windows p. 53
- perhaps heat mirrors in place of regular window glass p. 41

- make sure house is well insulated & weathertight, with energy-efficient features p. 29
- R-21–44 in ceiling pp. 31, 36
- R-12–28 in walls pp. 36, 37
- R-4–20 under first floor pp. 36, 39
- R-0–14 in foundation & basements pp. 36, 39
- double-glazed windows or storm windows, both with night insulation p. 40
- night-setback thermostats on heaters

- make sure house is well insulated & weathertight, with energy-efficient features p. 29
- R-21 in ceiling pp. 31, 36
- R-12 in walls pp. 36, 37
- R-4 under first floors pp. 36, 39
- R-0 in foundations & basements pp. 36, 39
- single-glazed windows adequate for most situations

Measure Your Resources

solar access:
- determine your insolation pp. 60–61, & orientation p. 64

money:
- the lower the number preceding the option, *usually* the lower the initial investment
- check your tax credit eligibility p. 265

skills & commitment: will determine if you buy or build, if you're prepared for the varying amounts of maintenance & repair necessary with different options

Options for Cold & Temperate Climates

Good Solar
full solar access & good clear skies during heating season

1. superinsulated design, especially in cold climates p. 33
1. direct-gain options such as ample south windows & clerestories; sunspace that can be closed off from living space, best with night insulation & added mass, particularly in high degree-day areas pp. 97, 110, 114
2. Trombe wall, or water or heavy-mass wall behind south glazing p. 112
2. solar-assisted heat pump pp. 20, 24–25
3. hybrid passive-active system p. 140
3. active system, especially effective in sunny, high degree-day climates pp. 137–40

any design will most probably need backup heat, such as electric resistance or fossil-fuel heating, or wood stoves

Intermittent Solar
partially shaded, less than full solar access, or partially cloudy during heating season, but reasonably good sun from 10 A.M.–4 P.M.

1. superinsulated design, especially in cold climates p. 53
1. direct-gain options such as ample south windows & clerestories; sunspace that has night insulation & added mass & can be closed off from living space pp. 97, 110, 114
2. Trombe wall, or water or heavy-mass wall behind south glazing p. 112
2. solar-assisted or conventional heat pump pp. 20, 24–25
3. hybrid passive-active system, particularly in low-temperature heat applications p. 140

any design will need backup heat, such as electric resistance or fossil-fuel heating, or wood stoves

Limited Solar
poor access, consistently cloudy skies, or heavy shade

1. superinsulation your best choice by far, especially in cold climates p. 53
2. double-glazed south windows with night insulation if any direct sun p. 40
2. heat pumps or high-efficiency conventional heating system pp. 20, 24–25
2. wood stoves p. 165

Options for Warm Climate
low-mass wood stoves p. 183
electric resistance heaters or other low-cost heaters, like kerosene, since they will be used infrequently pp. 11, 170
south windows that are well shaded for summer p. 123
sunspace, but night insulation and mass not necessary; need not be isolated from living space p. 114

You Are Buying/Building a New House

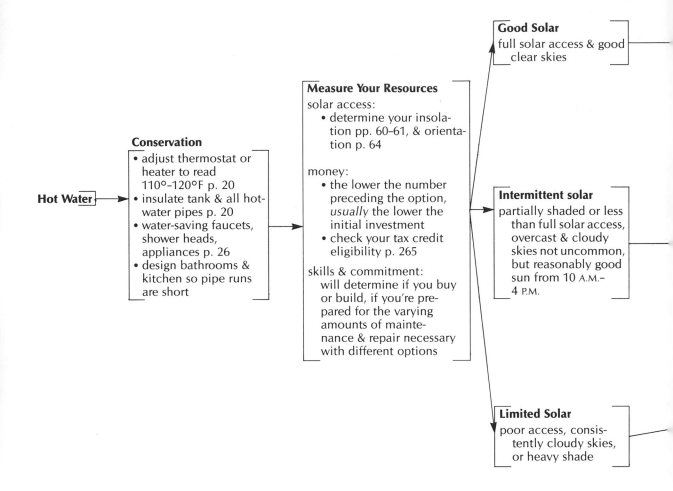

Hot Water

Conservation
- adjust thermostat or heater to read 110°–120°F p. 20
- insulate tank & all hot-water pipes p. 20
- water-saving faucets, shower heads, appliances p. 26
- design bathrooms & kitchen so pipe runs are short

Measure Your Resources
solar access:
- determine your insolation pp. 60–61, & orientation p. 64

money:
- the lower the number preceding the option, *usually* the lower the initial investment
- check your tax credit eligibility p. 265

skills & commitment:
will determine if you buy or build, if you're prepared for the varying amounts of maintenance & repair necessary with different options

Good Solar
full solar access & good clear skies

Intermittent solar
partially shaded or less than full solar access, overcast & cloudy skies not uncommon, but reasonably good sun from 10 A.M.– 4 P.M.

Limited Solar
poor access, consistently cloudy skies, or heavy shade

Options

1. batch heater; easiest owner-built option; good roof support critical because of weight, otherwise use ground mount pp. 62, 81
2. thermosiphoning heater with freeze protection pp. 62, 76
3. active heater with freeze protection pp. 62, 76
2. solar-assisted heat pump for hot water & space heating pp. 24–25
   pool and/or hot tub solar heating is probably possible in warm & temperate climates; pool cover recommended p. 82

any as preheater/booster to conventional water heating or point-of-use heaters p. 75

1. batch heater: easiest owner-built option, good roof support critical because of weight; otherwise use ground mount; consider insulating cover pp. 62, 81
2. thermosiphoning heater with freeze protection; consider insulating cover pp. 62, 76
3. active heater with freeze protection pp. 62, 76
2. solar-assisted or conventional heat pump for hot water & space heating pp. 24–25

as preheater/booster to conventional water heating or point-of-use heaters p. 75 whatever design you choose, pick a system that doesn't depend upon high temperatures to operate efficiently

1. conservation critical
1. tankless, point-of-use heaters p. 21
1. high-efficiency, well-insulated water heater with timer (for electric) or thermostat setback (for gas) to automatically turn down heater during no-use periods p. 27
1. conventional heat pump for hot water & space heating pp. 24–25

You Are Buying/Building a New House

Cooling

Conservation
- superinsulated or well-insulated design, as for space heating p. 53
- good attic ventilation (fan and/or roof vent) p. 44
- good cross-ventilation with operable windows on facing exterior walls p. 127
- window overhangs, awnings, trellises, shades p. 123
- vent and shade skylights, sunspaces pp. 46–47
- plant deciduous shade trees; vegetation rather than asphalt, stone, or concrete around house p. 123
- minimize exposed south windows in summer with reflective film or reflective insulation panels, especially in warm climates p. 41
- paint house and roof a light color, especially in warm climates p. 123

Measure Your Resources

money:
the lower the number preceding the options, *usually* the lower the initial investment

skills & commitment:
both will determine if you buy or build, if you're prepared for the varying amounts of maintenance & repair necessary with different options

Warm Climate
cooling degree-days
1,500–4,500 p. 260

Options
1. good insulated, weathertight house, or passive design; minimize dimensions of east & west sides of house & the window areas on east, west, & south sides (if such windows are critical for heating, make sure they are shaded); pay special attention to sufficient thermal mass to prevent sharp temperature rises during the day pp. 21, 53, 105
1. whole-house fan, thermal chimney, or overhead vents with good window & room layout designs for ample cross-ventilation & to bring cool air in from basement or shaded north windows or porch p. 44
1. heat pump may be more cost-effective heating & cooling system than separate conventional heating & conventional air conditioning pp. 24–25
3. if dry climate, evaporative or swamp coolers could be considered in place of conventional air conditioners p. 45

Temperate Climate
cooling degree-days
500–1,500 p. 260

1. good superinsulated or passive design; pay special attention to sufficient thermal mass to prevent sharp temperature rises during the day; provide south and west window exterior shading pp. 53, 105, 126–27
1. whole-house fan with good window & room layout for ample cross-ventilation & to bring cool air in from basement or shaded north windows or porch p. 44
1. heat pump may be more cost-effective heating & cooling system than separate conventional heating & conventional air conditioning pp. 24–25

You are Buying/Building a New House

Electricity

Conservation
- minimize electric usage for space heating & hot water whenever possible p. 44
- see cooling alternatives for air conditioning pp. 46–47
- install energy-efficient appliances & light fixtures p. 49; turn off when not in use
- if off-peak rates are available from your utility, install timers to automatically run dishwashers & dryers during off-peak times p. 47

Measure Your Resources

money:
- be aware that any wind or PV options will usually involve a greater initial investment than utility power will
- check your tax credit eligibility p. 265

skills and commitment:
- both will determine if you buy or build, if you're prepared for the varying amounts of maintenance & repair necessary with different options

solar:
   determine your insolation pp. 60–61, & orientation p. 64

wind:
- determine wind speed p. 212
- check for obstructions pp. 216–17
- check possible zoning restrictions p. 209

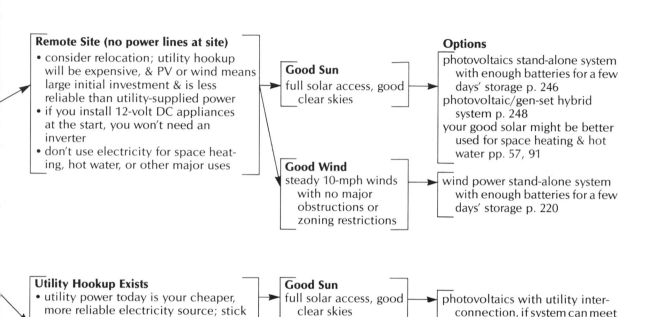

Remote Site (no power lines at site)
- consider relocation; utility hookup will be expensive, & PV or wind means large initial investment & is less reliable than utility-supplied power
- if you install 12-volt DC appliances at the start, you won't need an inverter
- don't use electricity for space heating, hot water, or other major uses

Good Sun
full solar access, good clear skies

Good Wind
steady 10-mph winds with no major obstructions or zoning restrictions

Options
photovoltaics stand-alone system with enough batteries for a few days' storage p. 246
photovoltaic/gen-set hybrid system p. 248
your good solar might be better used for space heating & hot water pp. 57, 91

wind power stand-alone system with enough batteries for a few days' storage p. 220

Utility Hookup Exists
- utility power today is your cheaper, more reliable electricity source; stick with it unless using renewable energy is more important than first cost & maintenance considerations

Good Sun
full solar access, good clear skies

Good Wind
steady 10-mph winds with no major obstructions or zoning restrictions

photovoltaics with utility inter-connection, if system can meet utility regulations p. 245
your good solar might be better used for space heating & hot water pp. 57, 91

wind power with utility inter-connection, if system can meet utility regulations p. 216

Conservation:
You Don't Pay for What You Don't Use

This conservation chapter is first in the book for two reasons:

- Energy saved is the cheapest energy
- Before you go solar, put in a wood stove, install a wind machine, or roof your house with solar cells, you should do everything possible to reduce your energy use by conservation so that your new energy system can run as efficiently as possible

To give you an example, here's what *Rodale's New Shelter* magazine (May/June 1981) learned when it tested five different types of solar water heaters for cost-effectiveness. The winner showed a return on investment of 19 percent, tax-free. But an $80 conservation project on an existing conventional water heater showed an almost unbelievable 321 percent return on its investment! That means it paid back its cost in less than four months—and continues to save the fortunate owners more than $200 a year.

Conservation savings won't be so dramatic for the whole house, but it will still be eminently worthwhile. The National Bureau of Standards surveyed 142 conservation retrofit projects on low- to middle-income houses in 12 cities. On average, upgrading just the structure of the houses cut energy bills by 17 percent. When heating systems were improved as well, energy use was cut more than 40 percent. Homes in Fargo, North Dakota, that already had attic and wall insulation plus storm windows saved an average of 40 percent on heating bills at a cost of just over $1,600 for additional conservation measures.

Some conservation projects are more cost-effective than others and it makes sense to do them first. Here's a list of those measures,

What You'll Learn in This Chapter

- Conservation can save you twice as much energy as solar heating can—at a lot less expense (page 3)

- Most of us inaccurately estimate the savings of conservation measures (page 3)

- You can do a good energy audit of your own home or hire one of the "doctors" who make house calls (pages 4–10)

- Why electric heating is generally the most expensive way to do the job, and how you can figure out what might be the cheapest for you (page 11)

- You can save money on water heating with relatively little effort (page 20)

- How a heat pump works and why it might be a better investment than a new furnace and air conditioner (pages 24–25)

- How night insulation— insulated shades, drapes, and shutters—plugs up one of the biggest heat losses in your house (page 40)

- How reflective films and heat mirrors for windows work hand in hand with your insulation to prevent heat losses (page 41)

- Why keeping summer heat out is so much more effective than getting rid of it after it gets in (page 44)

- How using off-peak energy can help you—and your utility company (page 47)

- About energy-saving appliance rip-offs and how to avoid them (page 48)

- If you're building a new house, a superinsulated house will probably save you more energy and more money than a solar design (page 53)

in order of significance, that generally give the greatest return on time and money invested:

1. Heating system improvement
2. Water heater improvement
3. Window insulation
4. Caulking
5. Weather stripping
6. Attic/ceiling insulation
7. Wall insulation
8. Floor insulation

These are general priorities, of course, and your situation may change their order somewhat. For example, if you've just installed a 95 percent efficient condensing furnace, you obviously don't need to spend any more money in that department. And if your floors aren't insulated at all and the rest of the house is, you should move floor insulation up from the bottom of the list.

Conservation Is Number One

Passive solar heating, one of the most effective uses of solar energy, may be more glamorous than conservation measures, but look at the relative contributions of passive design and conservation for residences in different cities in table 1-1. Passive heating has an appreciable impact on energy use, but if you have to choose one over the other, conservation is the obvious choice.

Your Residential Energy Budget

Surveys have proved that most of us don't really know which energy users in our houses cost the most, or where to focus our attempts to save money. For example, interviews with 200 Michigan homeowners showed that in only one category in seven did they accurately estimate energy savings resulting from conservation. They guessed right on adding insulation; it does result in the greatest potential energy savings. But they ranked lowering the thermostat number two, when in reality adding storm windows should be their second priority.

The unfortunate thing regarding misconceptions about which conservation measures do the most good is that we tend to follow through with those that are least effective and not give proper attention to the real energy savers. For a rough approximation, assume that an average household uses 60 percent of its energy dollars for space heating and cooling, 15 percent for water heating, another 15 percent for running appliances, and 5 percent each for cooking and refrigeration.

It makes no sense to spend a lot of time, planning, and money on lighting improvements to save $20 a year when a faulty furnace, poor insulation, and cracks and other openings are wasting almost 10 times that much. Situations differ, of course. For example, if you get all the free firewood you need and have no cooling needs in summer,

(Continued on page 10)

Table 1-1
Percentage of Total Energy Saving

City	Passive Heating (%)	Conservation (%)
Boston	26	74
Denver	41	59
Nashville	30	70

SOURCE: "Passive Principles: Conservation First," *Solar Age* (September 1981).

Home Heating Index

Your fuel bills contain essential information about your home's energy efficiency, but unfortunately that information is muddied by the fact that some winters are colder than others, that large homes usually need more heat than small ones, and other such truths. The following worksheet outlines a simple calculation that allows you to remove the "clutter" caused by these extraneous factors. The result, called the "Home Heating Index" (HHI), expresses your home's heating energy requirements in terms of Btus per degree-day per square foot of heated floor area (Btu/DD/ft²). This index can tell you whether your home is in good or not-so-good shape as a "keeper of the heat," and it can help you estimate how much you could expect to save by making various conservation improvements.

Recent research by the U.S. Department of Energy covering a large number of older homes in many different climates found that simple conservation measures such as weather stripping, caulking, attic and wall insulation, and furnace tune-ups could reduce the home's HHI to an acceptable eight. Thus, if your home's HHI is much higher than eight, conservation should be your first priority, and you ought to spend time here in Chapter 1 and then take a good look at the list of selected books for further reading at the chapter's end.

You will find yourself working with a few numbers in this worksheet, but be not dismayed. The directions take you step by step through what is really a simple process. A hand calculator would help, but even that isn't essential. You will have to dig up past energy bills, though if you can't, your fuel distributor or utility company will have a record of your account.

The energy use data you collect will have to cover at least one full year: for example, a period starting July 1, continuing through the fall, winter, and spring, and ending June 30 of the following year. Your bills must show the number of *fuel units* delivered (for example, gallons of oil, cubic feet of gas, etc.). Your result will be more accurate if you have several years' worth of fuel bills, provided you haven't made any major energy improvements or other changes in life-style during that time. For example; if 1½ years ago you increased your attic insulation from R-12 to R-30, you should not combine the fuel bills from the period before insulation with those from the period after insulation was added. (If you did want to estimate how much you saved by adding insulation, you would have to go back three years, to get *one full heating season* before insulation was added, and compare it to the most recent *full heating season after* insulation was added. As long as you didn't make other changes—such as adding a clock thermostat, lowering your thermostat settings, or getting a furnace tune-up—the comparison should be valid.)

The method presented here is a simplified version of a more detailed 22-page treatment given in *Solarizing Your Present Home* (Rodale Press, 1981).

Directions for Home Heating Index Worksheet

Note: In the worksheet that follows, a sample calculation is worked for an oil-heated home. Since oil is delivered by the tankful, and not continuously like gas through a pipe and electricity through a wire, the method of determining fuel use has some special requirements, which are described below. Propane is also a delivered fuel, so it is handled like oil. The methods for tallying gas, electricity, and wood use are also described, but beyond that step the worksheet is the same for all fuel types.

1. To add up your fuel use for one heating season, start with the first delivery (or meter reading) you received during or after the beginning of July, and ending with the last delivery you received during or before June of the next year. Add the column of *Fuel Units Used* to obtain the *Annual Fuel Use,* and enter this total on line ①A . *Fuel Units Used* refers to delivered amounts of oil or metered amounts of gas or electricity, which assumes that the fuel or power is actually used. Henceforth, the word "fuel" is meant to include electric power along with oil, gas, propane, and wood

2. If you used the same fuel for water-heating and/or cooking, you must make a *fuel use adjustment* to deduct this non-heating usage from the gross total you obtained in step ① . Choose a period during the summer when you had little or no heating load, and enter the number of days in that period on line ②E , the number of fuel units used in that period on line ②B , and the appropriate *summer conversion factor* in line ②C (found in table 1-2). In our example, it took 126.0 gallons to fill the tank on 8/25/78, which was 121 days after the previous fill on 4/26/78 (which is entered above the 8/25/78 fill date), so we enter 126.0 gallons on line ②B . Then you find the number of days between the *Previous delivery date* and the July delivery date (again, this is only for oil heating) and enter the number on line ②E . For electricity and gas you add up

(Continued on next page)

the fuel use indicated by the energy bills that came when there was no fuel use for space heating, or at least very little. Then you determine the number of days spanned by those bills.

If you use wood for space heating, you have to choose the MBtu per cord number from table 1-2 that best approximates the energy content of the fuel. Then you choose a seasonal efficiency number that best approximates the type and condition of your wood-burning appliance, and a summer conversion factor, both from table 1-2

3. Obtain *net annual fuel usage* on line ③Ⓐ by subtracting the *fuel use adjustment* (②Ⓐ) from *gross annual fuel usage* (①Ⓐ). The result in the example is 770.3 gallons per year

4. Convert fuel units to energy units by multiplying by factors ④Ⓑ and ④Ⓒ , obtained from the appropriate entry in table 1-2. In the example, oil contains 138,700 Btu per gallon and is used at 60 percent efficiency (0.6), which gives an annual energy consumption of 64.1 million Btu

5. Find out the *actual total degree-days* for the specific heating season for which you've entered your fuel use data. If you're studying more than one season, you'll need the degree-day totals for each one. Appendix A contains a degree-day map, but much more specific degree-day information can be obtained from your local weather bureau, fuel supplier, or electric utility company. In our example, the 1983-84 heating season had 5,400 degree-days, so we enter this number on line ⑤

6. Enter the *area of heated space* in your house on line ⑥ In our example, the house is 20 feet by 30 feet, two stories high, without a heated basement, so the area is 1,200 square feet (20 × 30 × 2 = 1,200)

7. Finally, obtain your *Home Heating Index* on line ⑦ by entering the values obtained previously from lines ④Ⓐ , ⑤, and ⑥, and dividing as shown

Armed with your HHI, look into table 1-3 to see where you stand and what improvements you can make to knock that number down.

Worksheet 1-1
Home Heating Index Worksheet/Calculations

(1) Tally of Fuel Use for One Heating Season

Your House		Examples (fuel oil)	
Date of Delivery or Meter Reading	**Fuel Units or Kwh Used**	**Date Delivered**	**Fuel Units Used**
		Previous Delivery Date (oil heating only) 4/26/78	—
		8/25/78	126.0 gal
		11/21/78	140.1 gal
		12/30/78	168.0 gal
		2/1/79	176.0 gal
		3/12/79	177.0 gal
		5/1/79	173.2 gal

(1A) = _____
Annual Fuel Use

960.3
gal/yr

(2) Fuel Use Adjustment

	summer fuel use	×	summer conversion factor	×	365 days/year	÷	number of days in summer period	=	fuel use adjustment
	(2B)	×	**(2C)**	×	**(2D)**	÷	**(2E)**	=	**(2A)**
Your house	_____	×	_____	×	365 days/yr	÷	_____ days	=	_____
Example	126.0 gal	×	0.5	×	365 days/yr	÷	121 days	=	190.0 gal/yr

(Continued on next page)

Worksheet 1-1—*Continued*

(3)

	gross annual fuel usage	−	adjustment	=	net annual fuel usage
	(1A)	−	**(2A)**	=	**(3A)**
Your house	_____	−	_____	=	_____
Example	960.3	−	190.0	=	770.3
	gal/yr		gal/yr		gal/yr

(4)

	net annual fuel usage	×	fuel energy content	×	seasonal efficiency	=	annual energy consumption
	(3A)	×	**(4B)**	×	**(4C)**	=	**(4A)**
Your house	_____	×	_____	×	_____	=	_____
Example	770.3	×	138,700	×	0.6	=	64,100,000
	gal/yr		Btu/gal				Btu/yr

(5)

Your house	$\dfrac{\text{degree-days}}{}$ =	
		actual total degree-days
Example	$\dfrac{\text{5,400 degree-days}}{}$ =	

(6)

Your house	$\dfrac{\text{ft}^2}{}$ =	
		area of heated space
Example	$\dfrac{\text{1,200 ft}^2}{}$ =	

(7)

	annual energy consumption	÷	actual total degree-days	÷	area of heated space	=	home heating index
	(4A)	÷	**(5)**	÷	**(6)**	=	**(7)**
Your house	_____	÷	_____	÷	_____	=	_____
Example	64,100,000	÷	5,400	÷	1,200	=	9.89
	Btu/yr		degree-days		ft²		Btu/DD/ft²

Table 1-2

Fuel Energy Content and Seasonal Efficiency Factors (for use in worksheet 1-1)

Fuel	Fuel Energy Content (use on line ④B of worksheet)	Seasonal Efficiency (use on line ④C of worksheet)	Summer Conversion Factor (use on line ②C of worksheet)
Fuel oil (#2)	138,700 Btu/gal	0.7 if system is new & has stack damper 0.6 if system is not new, but in good shape 0.5 if system is old & needs tune-up	0.5 for water-heating by "summer-winter" hookup with oil-fired space heating 0.9 otherwise
Natural gas	100,000 Btu/therm 103,500 Btu/100 ft³ 1,035 Btu/std ft³	0.8 if system is new & has automatic pilot & stack damper 0.7 if system is about average 0.5 if system is old & needs tune-up	0.9 (separate gas water heater)
Propane	100,000 Btu/therm 250,000 Btu/100 ft³ 2,500 Btu/std ft³ 91,500 Btu/gal	0.8 if system is new & has automatic pilot & stack damper 0.7 if system is about average 0.6 if system is old & needs tune-up	0.9 (separate propane water heater)
Electricity	3,414 Btu/kwh	0.95 for electric resistance 1.50 for heat pumps in climates with 4,000 DD or more 2.00 for heat pumps in climates with less than 4,000 DD	0.5 for conventional air conditioning 0.9 otherwise
Wood	19 MBtu/cord—dry hardwood 17 MBtu/cord—mixed 15 MBtu/cord—dry softwood	0.75 for top-of-the-line furnaces & catalytic wood stoves in good repair 0.50 for standard wood stoves 0.25 for all other wood-fired devices	0.5 for combined water/space heating 0.9 otherwise

SOURCE: "Do You Need More?" *Rodale's New Shelter* (November/December 1982).

(Continued on next page)

Table 1-3

How to Interpret Your Home Heating Index (HHI)

If Your Home Heating Index Is:	This Means That Your Home's Energy Efficiency Is:	What You Should Do:
Less than 4 Btu/DD/ft²	Excellent!	Recheck your calculations; suspect that a record of one or more fuel deliveries is missing. If calculation is OK, then relax! Adding insulation or weather stripping is probably not necessary & would probably not reduce your fuel bill significantly
4-10 Btu/DD/ft²	Better than average	You might succeed in reducing your fuel bills, but only if you can identify, by *careful inspection,* the exact locations where heat loss is occurring. Added insulation/weather stripping *may or may not* be worthwhile
10-20 Btu/DD/ft²	Average	You probably can reduce your fuel bills by adding insulation/weather stripping & tuning or upgrading your furnace. Carefully inspect your home to be sure you attend to the worst heat leaks first
More than 20 Btu/DD/ft²	Worse than average!	Recheck your calculations; suspect that one or more fuel deliveries was counted twice. If calculation is OK, then get to work! Almost any insulation/weather stripping you add will probably yield great savings. Do the least expensive retrofits first

SOURCE: "Do You Need More?" *Rodale's New Shelter* (November/December 1982).

priority number one, on the previous page, disappears and you can start with number two (unless wood heats your water, too).

At the direction of the federal government, utility companies will come to your house and check it for energy efficiency. Some companies will do this for free, while others charge a modest fee. Most homeowners using the service are pleased with the information they receive.

Trained, independent auditors, often called energy doctors because they specialize in diagnosing energy problems and prescribing cures, perform a service similar to that of the utilities, although generally in more detail. Besides checking your house, for an additional fee they'll do the work required to cut energy losses they find. Before calling in such a person, it might be a good idea to read this book through once—or at least this chapter, and especially the sample energy audit here—to be sure that the good doctor knows more about the subject than you do.

Some audits conducted by energy doctors take up to 22 hours, at a cost of more than $500. They involve an optical measuring technique called thermography that produces a color photograph showing where and how badly your house is leaking heat, and equipment to pressurize the house to determine the number of air changes it has per hour. A well-insulated house should have about one air change an hour, but leaky houses can have up to seven air changes an hour, thus wasting great amounts of warm (or cool) air.

The effectiveness of energy audits depends not only on how good the house doctor is, but on how much your house needs his services. For example, a Bonneville Power Authority survey of 18 houses in central Washington showed an average energy saving of just over 9 percent following an expensive private audit. The money saved on energy probably didn't pay for the audit. Since the houses were in very good energy-saving condition to begin with, an audit was probably unnecessary.

Maximizing Your Space- Heating Efficiency

On the average, Americans spend more than half of each household energy dollar on space heating and/or cooling. Those in the very cold North spend more on heating; those in Florida more on cooling. Your requirements depend on the climate you live in, but chances are good that heating your house is expensive enough to give you a strong incentive to do all you can to minimize its cost.

First off, are you heating with the cheapest fuel or energy source? Electricity costs vary with locality, and the all-electric home is no bargain if you have to heat it at 10 cents or more a kilowatt-hour. (However, if you have a heat pump, its high efficiency might just offset the high kilowatt-hour cost of that electricity. And spot heating with electricity, by using quartz heaters or radiant panels, might be the best way to go if you are heating isolated areas or looking for backup heat for your wood stove or solar system.) If natural gas is available, it may be your least expensive source of energy. Oil is also cheaper than most electricity. The price of wood varies from free if

(Continued on page 18)

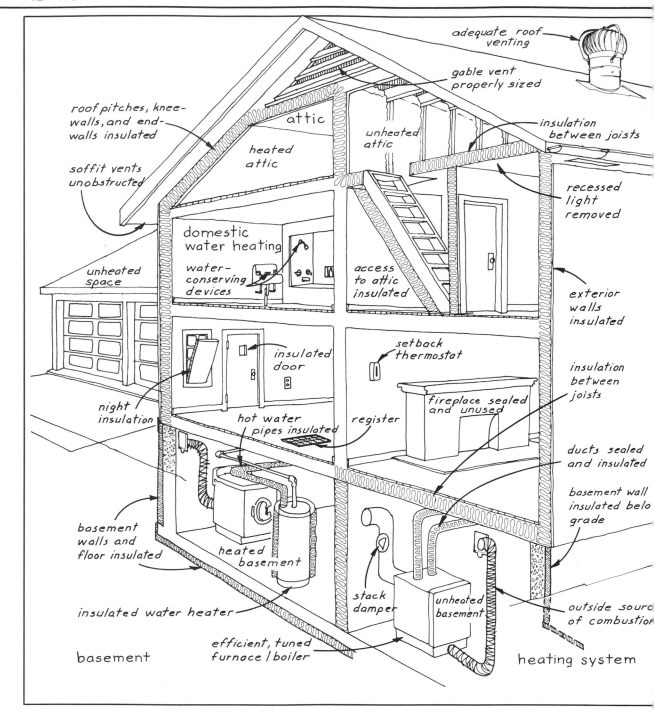

adequate roof venting

gable vent properly sized

insulation between joists

recessed light removed

roof pitches, knee-walls, and end-walls insulated

soffit vents unobstructed

attic

heated attic

unheated attic

domestic water heating

water-conserving devices

access to attic insulated

exterior walls insulated

unheated space

insulated door

setback thermostat

insulation between joists

night insulation

hot water pipes insulated

register

fireplace sealed and unused

ducts sealed and insulated

basement wall insulated below grade

basement walls and floor insulated

heated basement

stack damper

unheated basement

outside source of combustion

insulated water heater

basement

efficient, tuned furnace/boiler

heating system

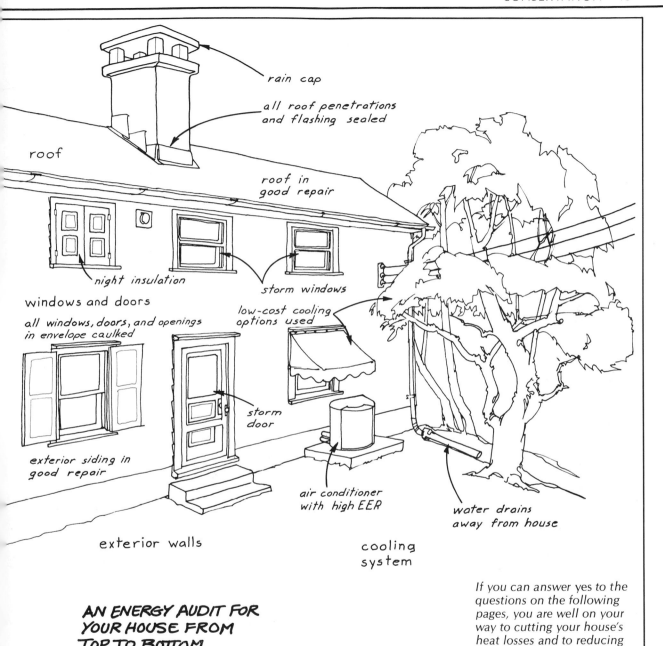

rain cap

all roof penetrations and flashing sealed

roof

roof in good repair

night insulation

windows and doors

storm windows

all windows, doors, and openings in envelope caulked

low-cost cooling options used

storm door

exterior siding in good repair

air conditioner with high EER

water drains away from house

exterior walls

cooling system

AN ENERGY AUDIT FOR YOUR HOUSE FROM TOP TO BOTTOM

If you can answer yes to the questions on the following pages, you are well on your way to cutting your house's heat losses and to reducing your heating, cooling, and electric bills.

An Energy Audit for Your House

Roof

Is the roof in good repair so leaks won't render insulation ineffective?

Does the chimney have a rain cap?

If the roof needs replacing, will you use light-colored shingles or roof covering?

Is there adequate roof venting to prevent moisture in the attic and/or rafters?

Are all penetrations in the roof sealed (chimney, vents, wires, and so forth)?

Is all roof flashing caulked?

Attic

Heated Attic

Do the roof pitches have insulation and a vapor barrier?

Are the kneewalls and endwalls insulated?

Unheated Attic (unoccupied)

Is there insulation (with a vapor barrier) between the floor joists?

Have recessed lights been removed and the space covered with insulation?

Is the doorway (or hatch) to the attic insulated and sealed?

Are the stairs, the sidewalls of the stairwell, and the door leading to the stairway insulated and sealed?

Is there a whole-house fan, and is it covered with insulation during the winter?

Are there gable vents to allow moisture to escape?

Are the eaves free from insulation so the soffit vents are not blocked?

Are all penetrations in the attic floor (for plumbing vents, ducts, chimneys, wires, and so forth) sealed with insulation?

Exterior Walls

Are the exterior walls insulated?

Is there a vapor barrier?

Is the exterior siding in good repair?

Are all cracks and openings in the building envelope caulked?

Are inlets and outlets for wiring, faucets, plumbing, and vents sealed?

Is the space between the sill plate and the foundation caulked?

Windows

Are there storm windows (except in the mildest climates)?

Is there caulk between the storm-window frame and the prime-window frame?

Are the prime windows caulked and weather-stripped?

Are the windowpanes secure in the framework?

Do all windows open easily and close tightly?

Does new construction include windows with heat mirrors?

Do the windows have night insulation?

Do the windows (south ones especially) have shading devices?

Are window air conditioners removed until summer or are they sealed and covered?

Are basement windows tight?

Exterior Doors

Are there storm doors on all exterior doors?

Are the doors weather-stripped?

Are the doors in good repair and do they close tightly?

Does the threshold fit tightly against the floor?

Is there caulking between the weatherstripping and the doorstop?

Does the door sweep fit tightly?

Are doors leading to unheated spaces (garages, workshops, and so forth) insulated on the unheated side?

Are basement doors tight?

Basement

Heated Basement

Are the walls insulated?

Are the doors and windows weather-stripped and caulked?

Do the windows have night insulation?

Is the floor insulated?

Unheated Basement (unoccupied)

Is there insulation under the first floor?

Are the walls insulated below grade?

Are windows, doors, vent openings, and plumbing and wiring openings sealed?

Are the spaces around pipes, wires, and ducts sealed with caulk or fiberglass insulation where they pass through the first floor?

Is the space between the wooden sill and the foundation caulked?

Are cracks in the walls sealed?

Does surface water drain away from the house?

Do drains around the house prevent water pressure from building up against the foundation?

Crawl Space

Is the crawl space insulated?

Are all sources of air infiltration sealed?

Is the crawl space adequately vented?

Concrete Slab (no basement)

Is the concrete slab insulated?

If no, is the perimeter of the slab insulated?

Heating System

Is your heating system efficient?

Does the flue pipe have a stack damper?

Do you regularly have the furnace or boiler tuned?

Do you change the furnace filter every six weeks?

Does the furnace or boiler have an outside source of combustion?

Is the heating system the proper size for the house?

Are forced-air ducts insulated and sealed with caulk or duct tape?

Are hot-water lines insulated?

Are registers or radiators free from obstructions?

Is there a setback thermostat?

Do you no longer use your fireplace?

If you have a wood stove, is it airtight and efficient?

Do you heat with solar energy?

Cooling System

See the cooling checklist later in this chapter

Domestic Water Heating

Are low-flow shower heads and low-flow aerators used?

Is the water heater insulated?

Is the temperature of the water heater set between 110°F and 120°F?

Does the water heater have a timer or thermostat setback that turns it off during parts of the day?

Are hot-water pipes insulated?

Is there an electric timer on an electric water heater and a thermostat setback on a gas water heater?

Is there a stack damper on a gas water heater?

Is there a solar domestic water-heating system?

Will you replace a worn-out water heater with a tankless one?

Scoring Your Home's Energy Efficiency for Heating

If the Home Heating Index you calculated on page 8 was a little or a lot too high, you're probably wondering how to get it lower with a few choice energy improvements. But which improvements should you choose? Worksheet 1-2 can help in answering that question. With a simple scoring system that covers six different areas of home heat loss, it can show you just how much better your home can be in its energy efficiency for space heating.

The first step is to add up the heating score of your house as it now is, including all energy improvements made to date. In the example, the scores listed under "Case A" refer to the house's as-is condition. The air infiltration score of 14 (line ①) indicates that several improvements have been made, according to the descriptions and scores listed in table 1-4. The attic/roof/ceiling insulation score of 10 (line ②) shows that there is an R-20 blanket over the top of the house. On line ③ the wall insulation score of 12 also shows that insulation has been added to get an R-12 insulation value. For windows, table 1-7 shows that a score of 12 (line ④) means the house has 10- to 30-year-old windows with storms, while a score of 6 for exterior doors (line ⑤) means that the original doors are still in use, but they are fully weather-stripped and covered by storm doors. In the basement/perimeter part of the worksheet, a score of 6 (line ⑥) relates to an unheated, uninsulated basement that contains no heating equipment. Thus, the as-is score adds up to a subtotal of 60, entered on line ⑦.

The next step is to determine the Heating Equipment Factor, which relates to the age of the heating system and to any improvements that may have been made to it. A factor of 1.6 for the example house, entered on line ⑧, shows a relatively efficient system that is kept in tune and that has a night-setback thermostat. This factor is a multiplier for the initial subtotal of heating scores. The final score of 96 (60 × 1.6) is essentially a base case score (line ⑨). As more energy improvements are made that score gets lower.

In the example, Case B lists the lower scores that relate to a variety of energy improvements in some of the six heat-loss categories. The new subtotal is 52, and the new Heating Equipment Factor is 1.5, which makes the final score 78. This number represents a 19 percent reduction in heating costs (96 − 78 = 18; 18 ÷ 96 = 19%, entered on line ⑩). Referring again back to the worksheet, the example house (Case A) used about 770 gallons of fuel oil for a season's worth of space heating. At $1 per gallon, the annual heating cost is $770. The 19 percent reduction means that heating costs are reduced by about $146 ($770 × 0.19 = $146).

You can make your own savings estimates by first plugging in scores for the as-is condition of your home, using the values listed in tables 1-4 through 1-10. Then you can use the Case B and C columns to look at different varieties or levels of conservation improvements. Bear in mind that this worksheet is really a ballpark estimator to help you see if you can make some cost-effective energy improvements beyond any you may have already made. In this game, you win with the lowest score.

Worksheet 1-2
Home Heating Scoresheet

Factor	Example		Your Home		
	Case A (house as-is)	Case B (house with some conservation improvements)	Case A (house as-is)	Case B (house with some conservation improvements)	Case C (house with more conservation improvements)
1. Air infiltration	14	10	_____	_____	_____
2. Attic/roof/ceiling	10	10	_____	_____	_____
3. Walls	12	12	_____	_____	_____
4. Windows	12	8	_____	_____	_____
5. Doors	6	6	_____	_____	_____
6. Basement/ perimeter	6	6	_____	_____	_____
7. Subtotal of items above	60	52	_____	_____	_____
8. Heating equipment factor	1.6	1.5	_____	_____	_____
9. Final score = ⑦ × ⑧ =	96	78	_____	_____	_____
10. % savings compared to base case		19%	_____	_____	_____
11. Current annual total fuel expense (example: 770 gal/yr @ $1/gal) =	$770		_____		
12. Estimated dollar savings due to conservation improvements ⑩ × ⑪ =		$146		_____	_____

(Continued on next page)

Table 1-4
Air Infiltration

Steps	If You Have: (check only one)	Then:
1	Done nothing	add 30
2	Installed caulking & weather stripping on all windows & doors	add 20
3	Done Step 2, plus caulked all around sill plate in basement, plus installed gaskets with caulking in all outlet boxes & light switches in outside walls	add 16
4	Done Step 3, plus caulked all pipe & wire penetrations in root of walls, plus weather-stripped all doors/hatchways into attic	add 14
5	Done Step 4, plus installed tight-sealing doors or dampers on kitchen fans & bathroom fans	add 10
6	Done Step 4, plus removed kitchen & bathroom fans & installed central air-to-air heat exchanger with ducts to kitchen & bath	add 8

Table 1-5
Attic/Roof/Ceiling

If You Have:	Then:
No attic/ceiling insulation	add 28
R-4–R-8 insulation	add 20
R-8–R-12 insulation	add 14
R-12–R-16 insulation	add 12
R-16–R-20 insulation	add 10
R-20–R-30 insulation	add 8
Over R-30 insulation	add 6

Table 1-6
Walls

If You Have:	Then:
No insulation in walls	add 24
R-4–R-8 insulation	add 16
R-8–R-12 insulation	add 12
R-12–R-16 insulation	add 10
R-16–R-20 insulation	add 8
Over R-20 insulation	add 6

Table 1-7
Windows

If You Have:	Then:
No storms; windows *more than* 30 yrs old	add 24
No storms; windows 10–30 yrs old	add 16
No storms; windows *less than* 10 yrs old	add 12
Storms; windows *more than* 30 yrs old	add 18
Storms; windows 10–30 yrs old	add 12
Storms; windows *less than* 10 yrs old	add 10
Any type of window with movable window insulation (e.g., pop-in thermal shutter, roll-down thermal curtain with edge seals), which is kept on window *all night* in winter	add 8

you cut it yourself to expensive if you buy it. Kerosene space heaters have become very popular; many folks find them to be an efficient, convenient way to heat a room or two at a time. (Check the latest reports of safety and emissions testing on these heaters before you invest in one, though; there is still controversy surrounding them.)

Table 1-8
Doors

If You Have:	Then:
Old doors *without* storm doors & *without* weather stripping	add 16
Old doors *without* storm doors but *with* new weather stripping & door sweeps	add 10
Old doors *with* storm doors & *with* new weather stripping & door sweeps	add 6
New doors & doorframes with magnetic-seal weather stripping plus storm doors	add 4

Table 1-9
Basement/Perimeter

If You Have:	Then:
Unheated, uninsulated basement with furnace & heating ducts/pipes in basement	add 10
Unheated, uninsulated basement with *no* furnace or heating ducts/pipes in basement	add 6
Basement walls insulated to R-8 or better (regardless of location of furnace)	add 4
Slab-on-grade foundation with *no* insulation	add 8
Slab-on-grade foundation with R-8 or better insulation	add 4

Table 1-10
Heating Equipment

Cases	If You Have:	Then Your "Heating Equipment Factor" Is:
1	Furnace/boiler *more than* 5 yrs old *without* yearly tune-up, service, & cleaning	2.0
2	Furnace/boiler *more than* 5 yrs old *with* yearly tune-up, service, & cleaning	1.9
3	Same as Case 2, *plus* night-setback thermostat in house	1.7
4	Furnace/boiler 2–5 yrs old *with* yearly tune-up, service, & cleaning, *plus* night-setback thermostat in house	1.6
5	Same as Case 4, *plus* stack damper	1.5
6	New flame-retention burner in furnace, *plus* night-setback thermostat & stack damper	1.4

New and Better Furnaces

If your furnace needs replacing, consider one of the new pulse-combustion or condensing furnaces. Furnaces 10 or 15 years old may be delivering only 50 percent of the energy in their fuel. The latest, most efficient furnace using conventional firing delivers about 84 percent. The new Lennox pulse unit goes as high as 96 percent. It doesn't need a chimney and it takes combustion air from outside, rather than from the living space, for higher efficiency.

Amana's new condensing furnace uses special heat exchangers that extract nearly all the heat from the fuel. Efficiency is 96 percent. There's no need for a chimney; exhaust gas is vented through polyvinyl chloride (PVC) pipe. And it's quieter than the Lennox.

Although a high-efficiency furnace costs considerably more than a conventional one, its fuel economy should pay back the difference in several years. For new-home installations, the elimination of a chimney alone might save the difference immediately.

Heat Pumps

A heat pump, which performs especially well in moderate climates, might suit you better than a new furnace. The drawing and box "Heat Pumps" explain the basic operation of a heat pump. One of the beauties of the heat pump is that it can be used for cooling as well as heating. In the cooling mode, refrigerant is pumped in the opposite direction to absorb heat from inside the house and give it up to the outside air.

Domestic Hot Water

Because you use hot water year round, your water heater is one of the biggest energy consumers in your house. Fortunately, it's also one of the easiest to save money on. Start by insulating it with a kit from your utility company or hardware store; such kits cost about $20 and are available for both gas and electric heaters. Also insulate the hot-water pipes to save even more heat; better a hot bath than a warm attic or basement.

It will also pay you to follow directions on the hot-water tank about draining off some water now and then to get rid of sediment, which slows the exchange of heat from the tank bottom to the water. If you have a gas heater, adjust the burner for proper combustion. If you have an old-fashioned pilot, see if you can upgrade the water heater to electric ignition, which fires up only on demand. This not only saves fuel, it's safer, too, because there's no open flame burning all the time.

One of the easiest and quickest ways to save on water heating is to turn the water heater thermostat down to 120°F or lower. You'll save money because your tank won't waste as much heat as it did at 140°F. It was this kind of conservation retrofit that *Rodale's New Shelter* magazine (March 1981) found paid for itself in about four months and saved more than $200 a year.

If your hot-water tank is a poorly insulated one (and most older models are), you can most likely improve its efficiency dramatically by wrapping it with fiberglass batts.

Tankless Heating

As much as 20 percent of the energy you use in heating water can be wasted because of heat loss from the tank and the long pipe runs between heater and points of use. *New Shelter* also found that annual standby heat loss from a 40-gallon electric water heater averages almost 600 kwh; that's $36 if you pay 6 cents per kwh, and $60 if you pay 10 cents.

If you have to replace your existing water heater, or are considering adding a bathroom, think about going tankless. Instead of heating a large tank of water in a central location, small tankless heaters do the job right at the point of use. They provide instant hot water and only as much as you need—none to waste heat in the pipes as it cools. Tankless heaters range from about $200 upward, depending on size. They're available in wall-mounted or under-the-sink models.

How Much Does Your Home Energy Really Cost?

Fuel Cost Comparison

Assumptions:

Natural gas
  Therm = 100,000 Btu = 100 ft³
  75% efficiency
  $/MBtu = 13.33 × $/therm

Fuel oil
  138,000 Btu/gal
  65% efficiency
  $/MBtu = 11.15 × $/gal

LP gas
  93,000 Btu/gal
  75% efficiency
  $/MBtu = 14.34 × $/gal

Electricity
  3,412 Btu/kwh
  100% efficiency
  $/MBtu = 293 × $/kwh

Mixed hardwoods
  24 MBtu/cord
  50% efficiency
  $/MBtu = $/cord ÷ 12

Mixed softwoods
  15 MBtu/cord
  50% efficiency
  $/MBtu = $/cord ÷ 7.5

Coal
  12,500 Btu/lb
  60% efficiency
  $/MBtu = $/cord ÷ 15

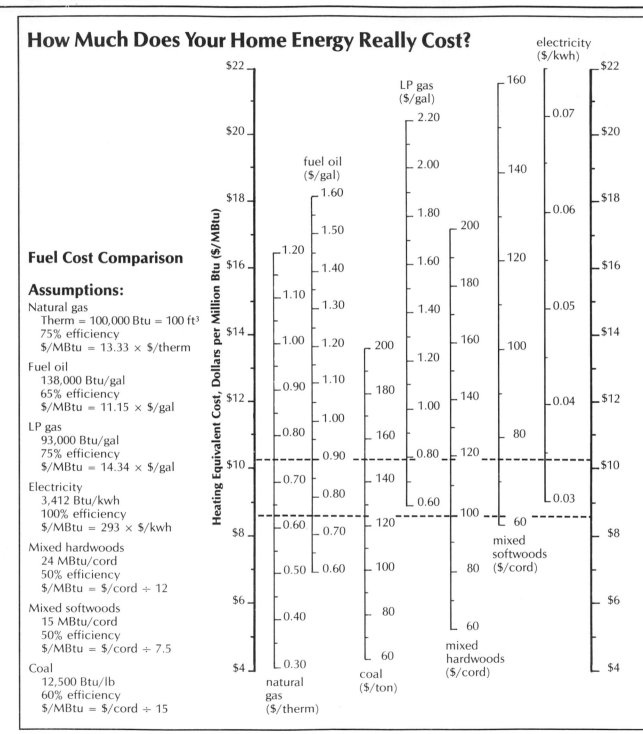

Today's ever-changing prices for different energy sources resemble a kind of horse race where everyone wishes they could put their money on the loser. For most people there's no bet because they're stuck with this or that fuel. But if you're in a position to change horses, you'll want to switch to the fuel with the lowest price. This nomograph gives you a way to compare energy sources with different unit costs (cents per kilowatt-hour, dollars per gallon, etc.). The standard unit is dollars per million Btu ($/MBtu), shown in the left-hand column. You can find the unit cost of the energy source you're interested in and then travel horizontally to the left to find the cost per MBtu. Finding the cost of energy in dollars per MBtu is good for comparing costs for different fuels, here, but it will also come in handy later on in other chapters. For example, you'll be using $/MBtu to find the energy-saving value of a solar water- or space-heating system and to compare the cost of wood with other heating fuels.

In this race, electric power is the perennial leader. Buying it at the national average of about 6¢/kwh is the same as buying fuel oil at about $1.55 per gallon or gas at about $1.30/therm. You're not likely to want to switch to electric heating from gas or oil, though you could make that switch if you were planning to use a heat pump. Standard resistance heating uses about 1 Btu of electricity to deliver 1 Btu of heat, but a heat pump system can give you a better deal: 2 Btu of heat delivered for every Btu of electricity consumed. That effectively cuts your cost for electric heating in half, which is the same as halving the cost per kwh. If you were spending 6¢/kwh for resistance heating, you'd be spending 3¢/kwh for heat pump heating. Now that's a price that in most places undercuts gas and oil. If you are planning a heating system for a new house, a high-efficiency heat pump could be your best bet (see pages 24 and 25).

Speaking of gas and oil, they have lately been neck and neck in the energy horse race. Prices for these fuels vary widely with location, and you'd have to look at local energy costs to find out which of the two is really the slower horse. Note that in the list of assumptions, oil is said to be used at 65 percent efficiency, while the efficiency of gas is 75 percent. Both of these numbers could be low with respect to the newest high-efficiency systems now on the market. Certain types of oil and gas furnaces and boilers have been tested at efficiencies as high as 96 percent. That has a big effect on the cost per MBtu, because that cost includes system efficiency. For example, oil at $1 a gallon used at 65 percent efficiency costs about $11/MBtu, but at 85 percent efficiency, the cost per MBtu drops to $8.50. So if heating systems were jockeys, you would do well to bet on the best one riding the slowest horse. The last shall be first (in energy savings).

To make sense out of the nomograph on the facing page, choose your fuel and then read down the column until you come to the price you pay per unit. Then travel right or left (they're both the same) to the heavier vertical line and you'll find your cost in dollars per MBtu. For instance, if you heat with oil and you pay 90 cents a gallon, your heating equivalent cost is $10 per MBtu. If, on the other hand, you keep warm with $100-a-cord wood, your heating equivalent is less, closer to $8.25 per MBtu. Redrawn from D. M. Stipanuk, "Comparing Heating Fuel Costs," (Cornell University, Cooperative Extension Northeast Regional Agricultural Engineering Service, 1979).

Heat Pumps

Some people are mystified when they try to understand how a heat pump operates. Heat pumps aren't quite magic, but they do pull off some fast-handed manipulation.

Inside every heat pump is a set of coils that contains a refrigerant, usually Freon. Part of the coil is located outside the house, and part is inside. Freon circulates continuously through the coil. On its way outdoors, the Freon passes through an expansion valve (see the illustration) that forces the Freon to "boil" into a gas. But the point is, it's a cold gas, a gas without heat, produced mechanically instead of by the means we're all familiar with: applying heat energy. This artificially produced gas thus has a heat deficit, and "wants" to absorb energy. That's precisely what happens when it hits the outdoor air. The outdoor air, cool as it may seem to you, is warmer than the cold gas, so some of its warmth is taken up.

Traveling back inside, the now-warm gas goes through a compressor that squeezes it into liquid form again. That intensifies the heat in the liquid and causes it to become artificially hot—hotter than the air surrounding the coil it's in. It "wants" to release heat now, which it does, to the surrounding air. At this point, a fan blows the released heat into your house. So heat pumps don't actually generate

HEAT PUMP CYCLE

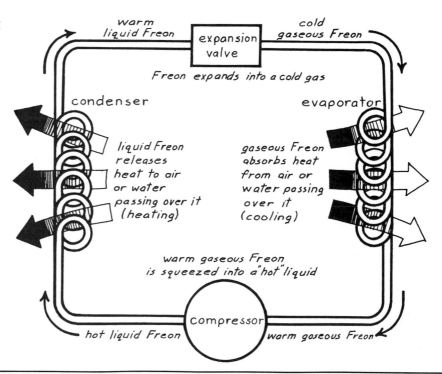

heat. They simply move it—pump it—from one place to another. The only electricity used is to run the compressor and fans.

In the summer, a reversing valve flips everything around, allowing you to cool your house and heat the outdoors.

Heat pumps are extremely efficient pieces of equipment. In most cases, you get twice as much energy out of a heat pump as you put in. Heat pump efficiency (in the heating mode) is measured by something called a coefficient of performance (COP). In simple terms, that's the ratio of the energy going into the unit to the amount coming out. If a heat pump has a COP of 2.5, it means you get $2.50 worth of heat for every $1 you put in. You just can't come close to that burning *any* kind of fuel, because there's no fuel that can give up more heat than it contains. The best you can do is a COP of 1. Electric resistance heat will give you that, but at great cost. Conventional natural gas furnaces have a COP of only 0.70, and oil burners usually log in at a measly 0.65. Of course, you have to remember that the power plants that generate the electricity in the first place have an efficiency of only about 30 percent. All said and done, though, heat pumps are still more efficient than oil or gas heat.

Of all the heat pump types, air-to-air heat pumps are by far the most common, since they're the cheapest to buy and the easiest to install. At their simplest, they are nothing more than reversible air conditioners that can be installed in a window. Even the largest takes up no more room than a gas or oil furnace. Retrofitting is an uncomplicated job, provided you already have a duct system to carry the heated air through the house. And for your initial investment, you get a heating system and central air conditioning in one package.

The investment, however, can be substantial. The average air-to-air heat pump costs about 25 percent more than a gas or oil furnace and an electric air conditioner, and that figure does not include installation and ductwork. Still, with savings of 25 to 50 percent on your annual heating bill, most buyers can expect a payback period of about four years. It all sounds pretty attractive, and in fact it is to many people.

But there are problems. For one thing, air-to-air heat pumps often aren't there when you need them most—when it's very cold. Although there are a few exceptions, most heat pumps can't handle air temperatures much below 35°F. It's about there that the refrigerant becomes too dense and the compressor can no longer work hard enough to pull out the heat that's in the air. The heat pump's COP drops, and the unit must automatically switch to a backup heat source to keep you warm. While there are a few pumps on the market that can use oil and gas as a backup fuel, in most cases the additional heat is provided by electric resistance strips inside the system.

Produce hot water where you use it and when you need it. Small tankless heaters, once found only overseas, can now be bought here. They provide hot water on demand, thanks to the gas or electricity that powers them.

Water Savers

Now that we've covered the water-heater end, let's look at the point of hot-water use. If you don't already have water-saving shower heads, try one (although I personally question their usefulness when dealing with teenagers who are going to use all the hot water no matter how long they have to stand there!). If a test convinces you that water-saver heads will work, put them in each bathroom.

Aerators for faucets mix air pressure with water pressure and therefore reduce the water flow rate without reducing the pressure. If your faucets don't have aerators, install them; they are cheap and effective.

How Heat Is Lost

After you've upgraded your heating system and hot-water supply, the next problem to tackle is that of heat loss through the shell of your house. First, let's look at the three culprits: convection, conduction, and radiation.

Water Conservation

Table 1-11 shows that there are some pretty big energy savings available for just a little bit of work in water-heating conservation. Most of the improvements you can make are low in cost; some are actually no-cost. The simplest is lowering the thermostat on the water heater, a no-cost job requiring just minutes but giving you substantial yearly savings. Added savings come with the addition of more insulation around the water heater. These savings appear in the footnotes of the table as reduced standby (heat) losses from the water heater itself. A cooler-running, better-insulated tank will naturally lose a lot less heat. In this case the standby losses are reduced from 20 percent of the total load to just 5 percent. Probably the next most cost-effective improvement is adding a low-flow shower head, which, as the table shows, saves a lot of gallons and a lot of energy. You can go as far as you like in this list of improvements and end up using perhaps 30 to 60 percent less energy for the same service and convenience you've always enjoyed.

To find the value of the energy savings, you simply multiply your cost in MBtu (taken from the box "How Much Does Your Home Energy Really Cost?" on pages 22 and 23) for water-heating energy (gas or electricity) times the MBtu/yr numbers under the "Savings" column. You'll probably find that the dollar savings are quite lucrative relative to the cost of making the improvements.

Table 1-11

Energy Savings from Water Conservation

Water Use (suggested conservation improvements)	Average Use (family of 4)			Conservative Use (family of 4)			Savings	
	Gal/Yr	Energy/Yr (MBtu)		Gal/Yr	Energy/Yr (MBtu)		MBtu/Yr	MBtu/Yr
		Electric*	Gas†		Electric‡	Gas§	Electric	Gas
Baths & showers (low-flow shower head; shorter showering times; eliminate most tub baths)	14,600	13.13	17.51	7,300	3.83	5.11	9.3	12.4
Clothes washing (warm instead of hot-water setting; fewer full loads per week instead of several partial loads)	4,015	3.61	4.82	1,460	0.77	1.02	2.84	3.8
Dishwashing (fewer full loads instead of several partial loads)	5,475	4.92	6.57	1,460	0.77	1.02	4.15	5.55

(Continued on next page)

Table 1-11—*Continued*

Water Use (suggested conservation improvements)	Average Use (family of 4)			Conservative Use (family of 4)			Savings	
	Gal/Yr	Energy/Yr (MBtu)		Gal/Yr	Energy/Yr (MBtu)		MBtu/Yr	MBtu/Yr
		Electric*	Gas†		Electric‡	Gas§	Electric	Gas
Food preparation (flow controls in faucets: restrictors, low-flow aerators)	1,095	0.99	1.31	730	0.38	0.51	0.61	0.8
Hand & face washing & shaving (no excessive running of water; faucet flow controls)	3,650	3.28	4.38	2,920	1.53	2.04	1.75	2.63
Household cleaning (use of warm instead of hot water; faucet flow controls)	730	0.66	0.88	730	0.38	0.51	0.28	0.5
Total yearly	29,565	26.59	35.47	14,600	7.66	10.21	18.93	25.68

SOURCE: Joe Carter, ed., *Solarizing Your Present Home* (Emmaus, Pa.: Rodale Press, 1981).

*Assumes 20% standby loss (energy lost through the walls of the water tank and pipes)
†Assumes 20% standby loss and 75% burner efficiency
‡Assumes 5% standby loss (15% lower than * and † because of tank insulation and thermostat setback)
§Assumes 5% standby loss plus 75% burner efficiency

Convection

Air rises by convection when displaced by cooler, denser air. Warmed air rises to the ceiling, or upstairs, and eventually into the attic and then outside. This is fine in summer when you want to get warm air out of the house, but it wastes money in winter.

Infiltration or drafts caused by wind also waste heat. If you can see daylight or feel air coming in around your doors or window frames, you have an infiltration problem. Cold air is replacing the warm air you've heated at great expense. Warm air is probably escaping through electric fixtures in walls and ceilings, too, and your losses are compounded when the wind blows. In summer, warm outside air infiltrates to replace expensive cool air inside your house. Conservation techniques for reducing air infiltration include caulking, weatherstripping, and just generally plugging up holes where heat can get in or out.

Conduction

Heat is also lost by conduction. As molecules of air, masonry, glass, or other materials are heated, they transfer heat to neighboring molecules. Eventually there's a steady flow of heat through the roof, foundation, walls, and glass from inside your warm house to the cold air or ground outside. In summer the problem is the reverse: heat from outside is conducted into the house. Nature tries to equalize temperatures, and the only way you can slow the process is to put a barrier between the warm interior and the cold outside. That barrier consists of appropriate insulation for all the surfaces mentioned above that protect the inside of your house from the outside environment.

Radiation

Radiation is the property of all materials to give off energy (heat) to the environment. All materials are radiating heat all the time. The sun radiates heat energy. Your outside walls, roof, and windows reradiate that heat. Again, the needed heat barrier is insulation or, in some cases, reflective material.

Stopping Air Leaks

Stopping air leaks by caulking and weather-stripping doors, windows, and other openings is generally cost-effective and an easy

This rather bizarre-looking photograph is called a thermogram, and it's designed to help you find the places in your house that radiate heat to the outdoors; in other words, leak heat. The light areas represent heat losses. Many home energy auditors can arrange to have a thermogram taken for you.

form of conservation. Adding insulation to your house can be cost-effective, but it's a much more costly conservation measure.

Caulking

Caulk door and window frames, electrical boxes, the corner joints of walls, where walls and foundation meet, wherever pipes enter the house, where the chimney meets the roof, around basement windows, and around clothes dryer vents and window air-conditioner units.

As you begin to caulk, you may find that you've got some holes that are just too big to fill with caulking compound. You'll know this when the caulking disappears as if into a bottomless pit. Put the caulking gun aside and plug the opening with something else first. Oakum, a treated hemp rope, can be tamped into big cracks. So can sponge rubber insulation or fiberglass. Even wood strips and rolls of heavy cloth can be used in plugging big gaps.

Once the hole is stuffed tight, seal it over with caulking. As a matter of fact, it's wise to seal over whatever you use with caulking to be sure you haven't left any gaps.

Check now for air leaks from living spaces into your attic, if you have one. These losses can be high if the attic isn't tight, and many attics are like sieves. Check for leaks around pipes and ducts that go into the attic, and check the fit of the attic door or door to the crawl space.

You may be losing lots of heat through fresh-air vents, stove hoods, and dryers. Air-to-air heat exchangers are available that get rid of humidity, pollutants, and odors, yet recover most of the heat from exhaust air.

Weather Stripping

Doors and windows are special problems because they must open and close as needed. The openings around them are great places for wind to whistle through and rob you of heat. This is where weather stripping comes in. It seals openings and "gives" to make for snug fits around windows and doors. Such material comes in a variety of types to take care of almost any application.

Insulating Your Home

It would be great to have so much insulation in your house that no heat could escape in winter or get in during hot weather. But insulation costs money and there are limits beyond which it isn't cost-effective to keep adding insulation. Table 1-12 shows the recommended R-values for different climates based on different heating degree-days. It will give you guidelines for appropriate insulation levels for your house. (Heating degree-days are the number of days times the number

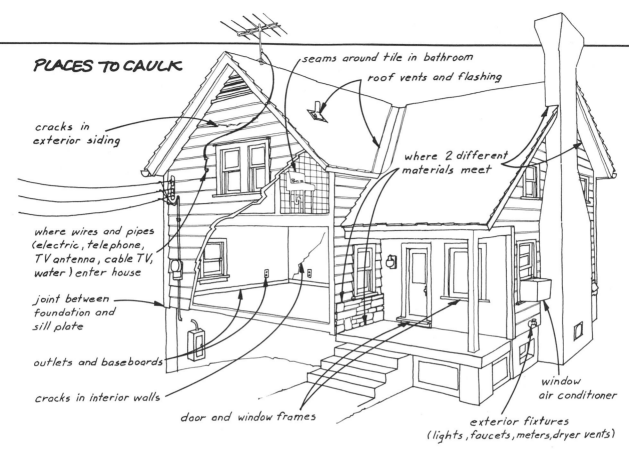

PLACES TO CAULK

- seams around tile in bathroom
- roof vents and flashing
- cracks in exterior siding
- where 2 different materials meet
- where wires and pipes (electric, telephone, TV antenna, cable TV, water) enter house
- joint between foundation and sill plate
- outlets and baseboards
- cracks in interior walls
- door and window frames
- window air conditioner
- exterior fixtures (lights, faucets, meters, dryer vents)

Even a reasonably well-insulated house may still be leaky if caulking isn't put in the right places. Compare your house to this one. In addition to the obvious cracks and holes, don't overlook the areas pinpointed here.

of degrees below 65°F during those days; see the map in Appendix A for rough estimates for your region.)

If you're fortunate enough to live in a relatively new house that was properly insulated in the first place, you may skim or skip this section and jump ahead to the discussion of night insulation for windows. Generally, the older the house, the less the insulation and the colder the occupants. If that describes you, read on.

Attic Insulation

The attic is usually the biggest heat loser, but it's also usually the easiest place to add insulation—and usually the most cost-effective, since heat rises and much of it can be lost through the roof.

There are essentially three approaches to insulating an attic, as shown on pages 34 and 35. If you don't use the attic for living space, the simplest solution is to install loose-fill insulation or fiberglass batts between ceiling joists and let the attic stay cold in winter. For a lived-in attic, add insulation between the roof rafters and endwall and kneewall studs to keep the attic warm.

850484

(Continued on page 37)

Weather Stripping for Doors and Windows: What the Experts Use

Material	Best Use*	Usually Made of . . .	
	Compression Gaskets		
Felt strip	Mail slots; storm windows; attic hatches; nonopening windows	Wool, hair, cotton, polyester	
	Open-Cell		
Foam tape	Attic hatches; storm windows; mail slots; exterior basement doors; pet doors	Polyurethane	
	Closed-Cell		
	Nonopening windows; wood casement windows; attic hatches; storm windows; mail slots; exterior basement doors; metal casement windows; double (dutch) doors; storm doors	Vinyl	
Sponge rubber strip	Attic hatches; nonopening windows; sides of standard doors	Neoprene, EPDM	
Gasket and flange	Pet doors; horizontal-slide windows: exterior basement doors; storm windows; metal casement windows; bottoms of standard doors; garage doors; nonopening windows; standard (double-hung) windows	Vinyl, rubber	
Reinforced gasket and flange	Storm doors; garage doors; double (dutch) doors; bottoms of standard doors; sides of standard doors	Vinyl lip or bulb; metal, wood, or plastic flange	
Bristles in retainer	*Sliding doors*; storm doors; horizontal-slide windows	Nylon pile bristle, metal or plastic retainer	

SOURCE: "Weatherstripping: Expert Choices," *Rodale's New Shelter* (January 1983).

*The applications in the "Best Use" column are listed in the order of preference given by the experts surveyed. Items in italic indicate that a majority (more than 50 percent) of the experts believe this is the best choice for a particular application.

Material	Best Use*	Usually Made of . . .

Tension Strips
Metal

Tension strip	*Sides of standard doors; metal casement windows;* standard (double-hung) windows; horizontal-slide windows; double (dutch) doors; exterior basement doors; wood casement windows; nonopening windows; sliding doors; storm doors; storm windows; attic hatches	Copper, brass, bronze, etc.

Plastic

	Standard (double-hung) windows; mail slots; wood casement windows; pet doors; sliding doors; horizontal-slide windows; metal casement windows; sides of standard doors	Polypropylene, vinyl, etc.

Specialty Designs

Door sweep	*Bottom of standard doors; garage doors*	Vinyl lip or bulb; metal, wood, or plastic retainer
Threshold gasket	Bottom of standard doors	Vinyl gasket; wood or metal threshold
Magnetic	Pet doors	Steel magnets and holders in plastic housing
Interlocking	Double (dutch) doors	Formed aluminum or brass
Astragal	Double (dutch) doors	Metal body, vinyl lip

Most of your home's heat is lost through the attic, but the attic is one of the easiest places to insulate. Keep in mind that you want to retain heat closest to your living areas. In an **unfinished attic,** this means insulating the attic floor. Install loose-fill insulation or fiberglass batts between the joists. Don't forget to insulate around the chimney and above the stair door. (If you have an attic ceiling fan, you can build an insulated box to cover it in winter, or insert a piece of fiberglass batt between the shutter and fan.)

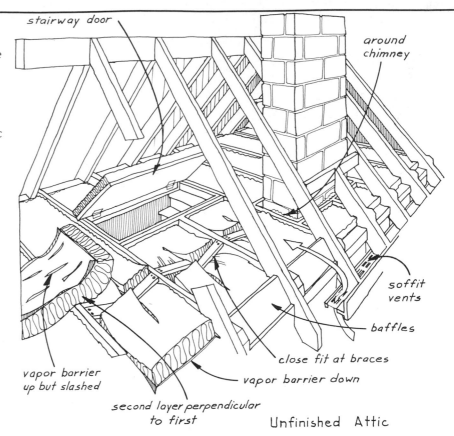

stairway door

around chimney

soffit vents

baffles

close fit at braces

vapor barrier down

vapor barrier up but slashed

second layer perpendicular to first

Unfinished Attic

ATTIC AND CEILING INSULATION

Vapor Barriers

You may be surprised to learn how much water vapor is released each day from various household activities. For example, a shower can vaporize about half a pint of water a day into the house; a large house plant may release three times that much. Cooking can release up to 3 quarts of water a day; washing clothes, about 2 quarts. And all this water can cause trouble.

When air is cooled sufficiently, any water vapor in it condenses back into water, and if this condensation takes place in your walls or roof, it can cause house problems like mildew, rotting, or paint failure. Condensation causes another problem of special concern in conservation: it can wet insulation, and when insulation is wet it isn't very effective. If you're in a geographic region where humidity is a problem, keep your insulation dry with a vapor barrier.

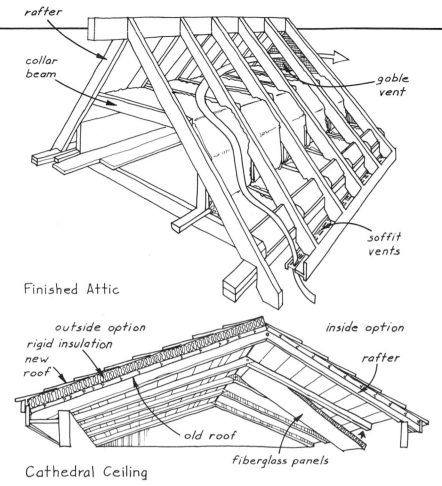

rafter

collar
beam

gable
vent

soffit
vents

Finished Attic

outside option

inside option

rigid insulation

new
roof

rafter

old roof

fiberglass panels

Cathedral Ceiling

In a **finished attic,** *install fiberglass batts between the rafters and the studs of the endwalls and kneewalls to keep in the heat. In either type of attic, be sure to allow space above the insulation for air to circulate.*

You have several insulating options for **cathedral ceilings.** *Placing rigid insulation on top of the existing roof, then installing new roofing above; or placing fiberglass panels inside, between the ceiling rafters, are two options.*

Wherever possible, install insulation with an attached vapor barrier of plastic material or aluminum foil. This barrier should face inside, unless you live in a region so very humid in summer that moisture can penetrate insulation from outside. In that case, the barrier should be outside. (Don't put them on both sides, however.) Vapor barriers not only keep out moisture but also reduce air infiltration by making the house tighter.

Sheets of clear plastic can be used as vapor barriers before installing insulation in floors or attics. Walls are often filled with loose insulation; here it's possible to add plastic vapor barriers on the inside wall and then put decorative paneling over it. A simpler and cheaper method is to use vinyl wall covering or paint formulated especially as a vapor barrier. Use an oil-base enamel first, and then add a coat of alkyd-base paint. A penetrating sealer or floor varnish can be used on wood floors or paneled walls.

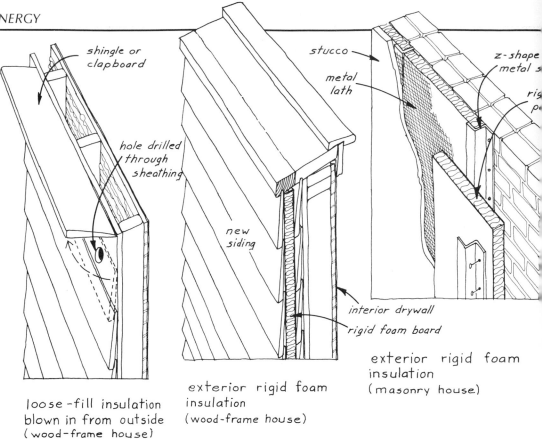

shingle or clapboard

hole drilled through sheathing

stucco

metal lath

z-shape metal s

rig
p

new siding

interior drywall

rigid foam board

exterior rigid foam insulation
(masonry house)

loose-fill insulation blown in from outside
(wood-frame house)

exterior rigid foam insulation
(wood-frame house)

WALL INSULATION : SOME OPTIONS

Table 1-12
Recommended R-Values

Heating Degree-Days	Attics and Ceilings		Walls		Floors over Unheated Areas		Foundations and Below-Grade Walls	
	Minimum	Preferred	Minimum	Preferred	Minimum	Preferred	Minimum	Preferred
Under 4,000	R-21	R-26	R-12	R-12	R-4	R-12	—	R-8
4,000–5,000	R-30	R-38	R-16	R-20	R-12	R-16	R-6	R-12
5,000–6,000	R-30	R-44	R-20	R-28	R-16	R-20	R-6	R-14
6,000–7,000	R-38	R-50	R-20	R-32	R-20	R-24	R-10	R-16
Over 7,000	R-38	R-60	R-28	R-40	R-20	R-30	R-12	R-20

SOURCE: Reprinted from "Do You Need More?" *Rodale's New Shelter* (November/December 1982).

NOTE: The values listed in this table refer to the approximate *total* R-value of the wall, floor, or ceiling, including insulation, sheathing, siding, wallboard, etc. In existing houses, it is almost always cost-effective to add insulation to achieve the "minimum" R-values. "Preferred" R-values represent a higher level of energy conservation that can sometimes be cost-effective for retrofitting in existing houses (depending on specific construction details) and is almost always cost-effective for building into new construction.

The values in this table are approximate guidelines, or rules of thumb. They are based on material such as the Residential Conservation Service Manual (DOE) and the California Energy Commission Standards but ultimately are our best judgments.

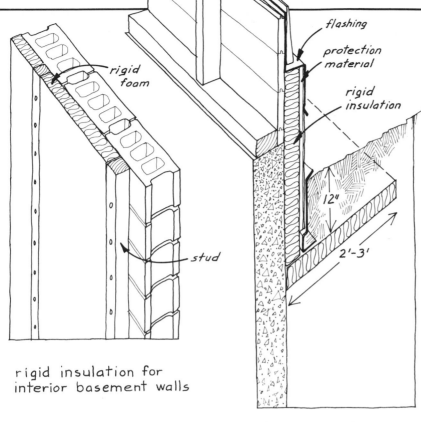

rigid foam

stud

rigid insulation for
interior basement walls

flashing

protection
material

rigid
insulation

12"

2'-3'

rigid insulation for
exterior basement walls

If the exterior and interior of a **wood-frame house** are in good repair and you don't wish to change the appearance of your house, a good insulation method is to blow in loose-fill insulation from the outside to fill the wall cavity. If the interior of the house is in good shape but the exterior needs work, you should consider adding new siding. Rigid foam insulation (polystyrene, urethane, or isocyanurate) can be installed **over the original siding,** and the new siding placed on that.

An increasingly common way to insulate masonry walls from the exterior is with rigid foam. In the method shown here, rigid polyurethane is friction-fitted between Z-shaped metal strips. Metal lath is screwed to the metal strips over the insulation, and stucco is applied over that. The **interior walls of a masonry house** can be insulated with rigid foam insulation and then covered by drywall. While this method (not shown here) preserves the exterior brick finish, any benefit from the brick as thermal mass is lost.

Interior basement walls can be insulated by building a frame wall and installing rigid foam insulation (shown here) or blanket insulation between the studs with a vapor barrier facing the living area. The insulation is then covered with drywall or paneling. Insulating a **basement from the exterior** requires the installation of two slabs of insulation. The horizontal slab prevents heat loss, which flows vertically from the basement.

If you have cathedral ceilings, you may want to use insulating board on top of the existing roof and add new roofing over that; or perhaps put the insulating board inside, under a new ceiling. This is a special case, and you should get professional help if you're not an expert yourself. Two types of insulation can be used: rigid insulation bonded to nailable sheathing, and rigid insulation of the kind usually used in walls. Bonded insulation includes NRG Nailboard, ThermaCal, and Thermasote. An R-value of 21 costs about $1.50 a square foot.

Unbonded insulation is cheaper, but you'll have to buy sheathing and install it separately. Nailing through foam insulation can affect its R-value, so again be sure to get professional help or advice. Typical commercial products include High-R, Styrofoam TG (tongue and groove), Thermatite Plus, and Thermax.

Wall Insulation

Next come the walls. Check the type walls you have, and find out what kind and how much insulation is in them. R-13 works out

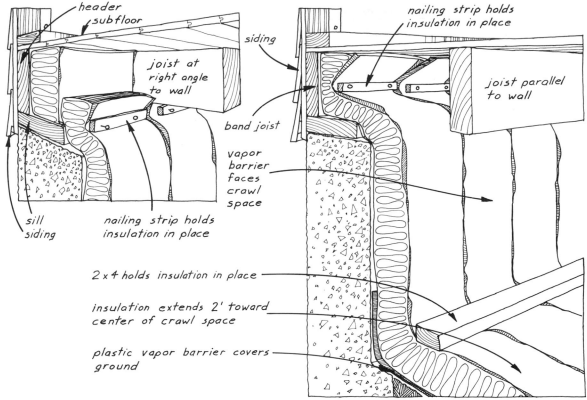

Labels in figure:
header
subfloor
joist at right angle to wall
siding
band joist
vapor barrier faces crawl space
nailing strip holds insulation in place
joist parallel to wall
sill siding
nailing strip holds insulation in place
2 x 4 holds insulation in place
insulation extends 2' toward center of crawl space
plastic vapor barrier covers ground

INSULATING A CRAWL SPACE WITH BLANKET INSULATION (2 options)

There are two ways to **insulate crawl spaces** with blanket insulation. First, in crawl spaces where joists are at right angles to the wall, install short pieces of insulation between the joists and tightly against the header. Longer pieces are attached to the sill and extend down the wall and 2 feet into the center of the crawl space. Second, in a crawl space where the joists are parallel to the wall, blanket insulation is nailed to the band joist. The insulation extends down the wall and toward the center of the crawl space, also for 2 feet. In both cases, a plastic vapor barrier covers the ground.

to 4 inches of fiberglass, rock wool, or cellulose. If you have a 4-inch stud wall filled with any of them, or an 8-inch block wall filled with vermiculite, you're close enough and can go on to windows. If not, stay with us.

Loose-fill insulating materials can be blown into uninsulated walls. It requires lots of hole drilling and plugging, but is often easier than stripping off the siding or interior wallboard so that batts or roll insulation can be installed. Another option is to add a layer of rigid foam insulation (polystyrene, urethane, or isocyanurate) either inside or outside the wall. Rigid polystyrene insulation is popular for this application, and each inch adds R-3.5. Use closed-cell foam to resist moisture. A decorative and protective coat of stucco may be applied to an outside layer of rigid foam insulation, and it adds more insulation as well. Gypsum board can go over the rigid foam insulation if you're insulating from inside.

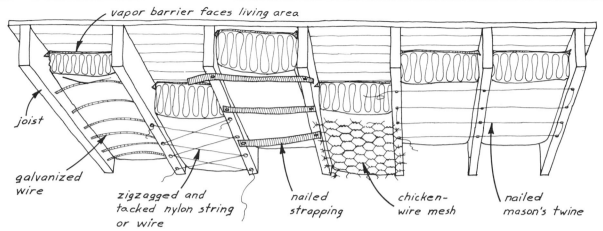

vapor barrier faces living area

joist

galvanized wire

zigzagged and tacked nylon string or wire

nailed strapping

chicken-wire mesh

nailed mason's twine

FIBERGLASS BATTS INSTALLED UNDER FLOOR
(5 options to hold insulation in place)

Aluminum foil is also a good insulating material. A single sheet of bright foil suspended in a hollow wall increases the R-value from less than 1 to about 2.6. Adding several sheets increases the effectiveness, making use of trapped air. Three sheets of foil in a wall cavity (creating four air spaces) gives about R-11, equal to 3½ inches of fiberglass. A product called Foilpleat uses three attached sheets of foil.

Floor and Foundation Insulation

With ceiling, attic, and exterior walls properly insulated, you have blocked most of the heat losses. If money has run out, you've spent what you had in the right places. To finish the job in style, however, check the floor insulation in your house. Let's start with a concrete slab floor. Ideally, it should have been poured over polystyrene foam slabs. If it rests right on the ground, the only good way to insulate is to build up the floor—a major job that I doubt you'd find worth the cost and effort.

What you have to do is lay down a plastic vapor barrier, a layer of Styrofoam slabs, furring strips, and finally a plywood subfloor to prevent crushing the foam. Finish off with flooring or carpet. A much simpler, though less effective, approach is to lay down wall-to-wall carpet with an insulating pad underneath.

A floor above a heated basement doesn't need insulation, of course. But if your floor, or part of it, is over an unheated basement, you should add insulation between the floor joists. Fiberglass batts are easily added; use enough to give the recommended R-value for floors. If the floor is over a crawl space, add insulation to the walls and the nearby ground of the crawl space as shown opposite.

If the heating system and water heater are not in the basement of a house, and if the basement is not used as living space, **the first floor** *(instead of the basement walls)* **can be insulated** *by installing blanket insulation between the joists. Shown here are five ways to hold the insulation in place.*

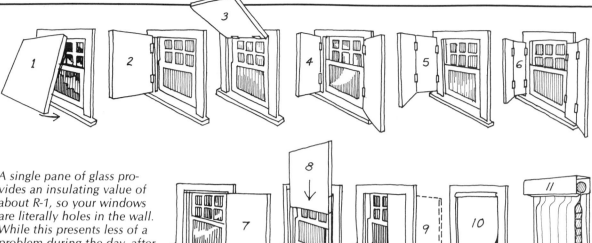

INTERIOR NIGHT INSULATION

A single pane of glass provides an insulating value of about R-1, so your windows are literally holes in the wall. While this presents less of a problem during the day, after sunset the heat rapidly escapes from your home. Interior shutters and shades keep the warmth inside.

Exterior shutters are not as convenient to use as interior ones, but for large windows, they eliminate the problem of interior shutter storage space. Some exterior shutters have reflective inside surfaces so that when they're opened, the sun reflects off this surface into the window to increase solar gain in winter.

Interior:
1. pop-in panel
2. side-hinged shutter
3. top-hinged shutter
4. double shutter
5. bifold shutter
6. double bifold shutter
7. horizontally sliding shutter
8. vertically sliding shutter
9. self-storing shutter
10. roman shade
11. thermal drapes with top valence

Exterior:
1. bottom-hinged shutter
2. double shutter
3. double bifold shutter
4. top-hinged shutter

Windows—A Special Problem

Your house is now properly insulated from top to bottom and all around the sides except for one important area—the windows and glass doors. These transparent openings are the remaining chinks in the insulating armor. Even though you've caulked and weather-stripped around their edges, your windows and glass doors are losing far more heat than the walls.

You may have R-13 insulation in your walls and floor, and R-26 in your ceiling. But a single pane of glass provides only about R-1! So 13 times as much heat per square foot escapes through the glass as through the walls.

Perhaps you already have double-glazed windows and storm doors. The extra layer of glass boosts the value to about R-2, which means it saves about half the heat you lost through R-1 windows. Triple glazing, recommended only for very cold areas, gives about R-3 insulation. Three panes of glass is the maximum. Glass is expensive, and more than three panes will cost more money than you'll save in energy.

The simplest additional glazing is plastic film, temporarily but securely attached to window and door frames with plastic tape. This adds insulation by trapping a layer of air between it and the glass, and also reduces the heat loss caused by infiltration.

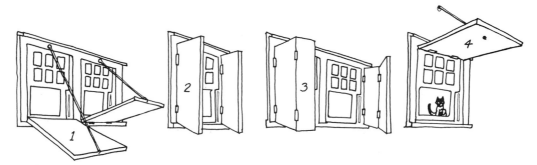

EXTERIOR NIGHT INSULATION

Heat Mirrors

New on the market are heat mirrors, special clear films that are far better one-way heat traps than ordinary windows. Special windows are fabricated with the heat mirror film between two normal panes of glass to create a glazing sandwich with an R-value 1½ times that of double glazing, and half again that of triple glazing. This application is probably more cost-effective on new-house construction, and you will have to do some figuring to decide whether or not to use it as a retrofit.

Reflective Films

Mirrorlike plastic films have long been used to reflect the sun's heat from windows in commercial buildings. Now they're available for homes. But don't expect reflective film to work both ways at once; it can't let solar heat in during the winter and keep it out in summer. Some films can be removed, which makes it possible for you to use the film in your house in summer by applying it on the outside of your windows, and reverse the process in winter to keep inside heat from getting out. Some films are applied wet; others have a self-sticking adhesive backing. In all cases, follow the manufacturer's directions carefully. Some films are not guaranteed unless professionally installed.

Night Insulation

After sunset, the insulation problem becomes much greater as heat moves only one way through your windows—from warm interior to cold outside. The only way to do anything about this heat loss is to cover windows with insulation at night.

Many people think that curtains or drapes provide good insulation. Curtains do help by trapping air between themselves and windows

(Continued on page 44)

Movable insulation—such as insulated shades, quilts, foam beads, and interior and exterior shutters—reduces nighttime heat loss through glazed areas by cutting drafts along window edges and by putting extra insulation right over the glass itself.

and thus reducing air movement. But tests show that ceiling-to-floor, boxed drapes cut heat losses by only about 20 percent. If they cover just the window, and don't seal at top and bottom, they save only 10 percent.

There are a number of window insulation products on the market from sew-them-yourself roman blinds to decorator thermal blinds or shutters with good R-values and a tight-sealing track along the edge of the window. Insulating shutters or panels of foam offer even higher R-values. Any air leaks at the edges reduce their effectiveness, however, so they must fit tightly. Where space is available, large panels can be hinged at the top and swung down over the windows at night.

Panels of high R-value insulation sliding horizontally in snug-fitting tracks are excellent. If there's ample wall space beyond the window, a one-piece panel eliminates the center joint where two panels would meet. Ends of the panel should be held tight to the wall to prevent air leaks. Two-inch slabs of Styrofoam are thin enough to be unobtrusive, especially if curtains or drapes are drawn over them, and provide about R-10. Two inches of urethane foam equals R-14; added to the window itself, this about equals an R-19 wall.

Rigid foam panels can be used *between* existing sliding glass doors and their storm doors for winter-long insulation. With some ingenuity, large R-2 glass areas can be upgraded quickly and cheaply in this way.

Be sure that the insulating materials you use aren't flammable, toxic, or otherwise dangerous. Check building codes before you use a seemingly superior material that may be hazardous.

Exterior Shutters

In early American architecture, exterior window shutters were almost standard features and provided extra protection against the weather. They made sense then, and if you use good insulation, they can make even more sense today. Such shutters reduce infiltration heat losses and are better at keeping the summer sun out. They can also be lined with reflective material and partially opened in winter to reflect more solar heat into the house.

Exterior shutters have their drawbacks, though. You must go outside to open or close them, and you might forget once in a while, or get lazy about the constant chore. Shutters exposed to an outside environment also need adequate protection from the elements.

Summer Cooling

Remember that most of what you do to cut heating costs also cuts cooling costs during the heat of summer. Make sure you've taken care of heat losses by plugging leaks, insulating properly, and using reflective films, screens, or awnings over glass exposed to the direct sun.

The best way to keep your house cool when you want it cool is to block solar heat at the point of entry. What can't get in can't make you miserable. Well-insulated, light-colored roof and walls, plus awnings or other shade devices and perhaps some reflective film on windows, will keep your air conditioner from running so often.

As much as possible, run your heat-producing appliances at night instead of during the day. Be sure the furnace pilot light is off during summer because it wastes energy two ways: the pilot burns expensive fuel, and its heat adds to the cooling load and contributes nothing of benefit.

Natural Ventilation

As much as you can, work with nature and not against it. Just as you open your house to the warm sunshine in winter, open it to the cool air and skies on summer nights. And don't neglect the cooling effect of cross-ventilation when air outside is at least as cool as it is inside. Whole-house fans are deservedly popular for their ability to sweep in cool outside air and flush out hot air.

Air Conditioning

Refrigerated air conditioning is very expensive, both in operation and in maintenance costs. If you have to replace an air conditioner, look for one with a higher energy efficiency rating (EER) and a fan-type precooler for the condenser. Consider heat pumps, too; they're usually more efficient than all but the newest furnaces, and can beat the performance of conventional air conditioners, too. For more about them, turn back to pages 24 and 25.

Evaporative Cooling

If yours is a dry, hot climate, you may choose evaporative cooling over refrigerated air conditioning. A wet cloth hung in an open window exposed to a breeze begins to dry. As the water evaporates, heat is absorbed and the air—and people nearby—are cooled. This simple process is the basis for evaporative cooling. Even the old swamp coolers do a creditable job in relatively dry environments; newer designs are better. Two-stage evaporative coolers provide dry cooling at a fraction of the cost of refrigerated air conditioning. Some manu-facturers offer "piggyback" units, in which an economical evaporative cooler operates during dry weather and refrigeration cooling takes over only when it gets muggy.

Evaporative coolers are mounted on roofs or walls and a fan or blower pulls outside air through water-soaked pads, cooling it before

Be Cool: A Checklist of Cooling Options

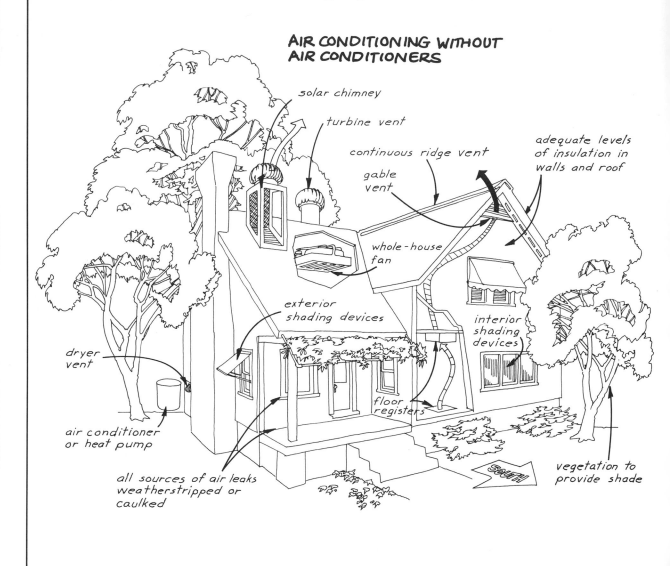

AIR CONDITIONING WITHOUT AIR CONDITIONERS

solar chimney

turbine vent

continuous ridge vent

gable vent

adequate levels of insulation in walls and roof

whole-house fan

exterior shading devices

interior shading devices

dryer vent

air conditioner or heat pump

floor registers

all sources of air leaks weatherstripped or caulked

SOUTH!

vegetation to provide shade

it goes into the house. They have one big drawback; they add moisture to the air. When the saturation point is reached, you might as well turn the evaporative cooler off, or at least shut off the water to the pads and rely on air movement only.

No-Cost to Moderate-Cost Options

Wear lightweight, loose clothing

Wear sandals or beach thongs, or no shoes at all

Don't open the refrigerator door frequently

Plan outdoor activities to prevent heat gain in house

Replace an old refrigerator with an energy-efficient one

Cook outdoors on a grill or solar cooker

Use heat-producing appliances at night instead of in the heat of day

Use incandescent lighting only when necessary

Turn off the pilot light on the furnace

Vent the dryer to the outside

Install an exhaust fan in the bathroom and kitchen

Hang clothes outside to dry

In humid climates, keep swimming pool near the house covered

Install a continuous ridge vent on the roof

Install vents in the roof

Use exterior shading devices (especially for east and west windows): awnings, overhangs, shutters, screens, canopies, trellises

Use interior shading devices: drapes, screens, reflective film, shades, blinds

Plant arbors, hedges, deciduous trees, vines, and so forth to provide shade and cooling

Replace an old, worn-out roof with a light-colored one

When repainting your house, choose a light color

Use portable, boxlike fans to promote air movement

Use ceiling fans

Keep windows and doors shut during the day

Open windows and doors at night to allow breezes in

Open windows at house's lowest point and at the highest point to create natural air movement

Install a whole-house fan

Install a solar chimney as an alternative to a whole-house fan

Install floor registers so that as hot air leaves the gable vent, cool air from the basement or crawl space is drawn into the house

Weatherize your house with weather stripping, caulking, and adequate amounts of insulation in the ceilings, floors, roof, and exterior walls

Most Costly Options

Evaporative coolers (for dry climates)

Heat pumps and air conditioners with high EERs (located in the shade)

Roof-sprinkling systems (for dry climates)

Installation of mass to increase a house's ability to store coolth (such as water-filled containers, tile, brick, concrete)

Off-Peak Energy Use

In the dream world of utility executives, consumers would use the same amount of electricity all the time and put a steady load on the power plant around the clock. In real life, however, there are times of very little energy use and also nightmares called peaks that sometimes turn into brownouts and blackouts.

In an effort to even out this fluctuating power load, some utilities offer lower rates during off-peak hours to encourage customers to use electricity during times of low demand. See if this is the case where you live, and try to take advantage of cheaper electricity. Timers and other hardware are available so that you can wash and dry the clothes and dishes in off-hours instead of when everyone else is loading the power grid. There are even electric resistance heaters that store heat during off-peak hours and release it gradually during the day.

The Great Energy Rip-Off

It's fairly well known by now that some washing machines, TVs, refrigerators, furnaces, and air conditioners use less energy than some others. EER labels on appliances indicate how much electric power can be saved. Of course, a more efficient appliance generally costs more to manufacture and thus more to buy.

Historically, consumer indifference has been most to blame for the poorly insulated houses and wasteful appliances we're stuck with. But times are changing, and people are asking for better energy performance in things they buy. Manufacturers have picked up on this interest and new products can bring you real energy savings, as you've already seen in this chapter. Unfortunately, however, some "energy-saving" products don't perform anywhere near their manufacturer's claims.

The federal government's General Accounting Office has reported a number of consumer rip-offs thriving on energy crisis concerns. Such occurrences jumped fourfold between 1975 and 1980 and are continuing. You may have been tempted—or even taken in—by some of them. Here are samples:

Ceiling fans were found to save not 50 percent on heating bills but closer to 5 to 10 percent; in houses with ceilings less than 8 feet high, the heating savings with ceiling fans is almost zero

Furnace dampers advertised as saving 30 percent on fuel turned out to save only about 8 percent on oil furnaces and 5 percent on gas furnaces

House siding promoted as saving up to 40 percent was actually found to save less than 5 percent

Furnace flue heat-recovery devices (costing about $500) saved only 7 percent in heating costs instead of the 50 percent claimed

The National Aeronautics and Space Administration (NASA) made a valiant effort to use some of its technology to develop a device called a power factor controller for residential use. NASA's invention does work with industrial motors that are used constantly by monitoring the load and supplying only as much electric power as is needed. A number of private companies manufactured smaller versions for residential use in washing machines, refrigerators, dryers, air conditioners, and other appliances. Selling for as much as $50, the controllers were advertised as saving up to 60 percent of the energy normally used. Tens of thousands were sold. Unfortunately, subsequent tests have proved that savings are nowhere near 60 percent. In fact, savings are so small they'll probably never pay back the cost of the devices.

Two types of devices called light buttons and advertised to save electricity use in light bulbs seem to be a waste of money, too. Costing from $2 to $4, they also reduce light output. One type is also a potential fire hazard because of its high operating temperature. You might just as well substitute a smaller bulb.

More recently, the Federal Trade Commission ordered a company to stop advertising that its plastic film storm windows provided an insulating value of more than R-9. Laboratory tests showed that the actual value was less than R-2.

There are well-meaning exaggerations in every industry, and conservation is no exception. But there are also some blue suede shoes in this business, so be alert for claims that sound too good to be true. They generally are.

Appliances

While appliances don't use as much energy as heating and cooling do, they use what is generally the most expensive form of energy—electricity. Table 1-13 shows the amounts of electricity used by major and small appliances. Knowing what they cost will make it easier to set up some rules for budgeting their use.

Here are some general buyer and user tips on a variety of appliances. See how many you're already using.

Refrigerator

Check EER ratings on new ones for comparative efficiency (although EER ratings are not compulsory, many manufacturers use them)

Separate doors for freezer and refrigerator save energy

A self-defrosting model may use 50 percent more energy; if you've already bought one, turn off the control and save electricity

Top or bottom freezer is more efficient than side-by-side

Coils behind refrigerator should be cleaned regularly

Built-in ice-makers and water coolers are expensive to operate

Don't put refrigerator in warm part of room

Door should seal tightly; check gasket regularly and replace if loose

Don't use too low a temperature setting

Power saver switch available on some refrigerators saves about 15 percent

Leaving the refrigerator door open 12 seconds means a complete air change and wasted energy

Don't pack food too tightly; allow for air circulation around food inside refrigerator

Range and Oven

Use electric ignition on gas range if possible—pilots waste gas

The self-cleaning feature uses lots of energy

Opening oven door for even a few seconds drops oven temperature 25 percent

A microwave oven uses 30 to 70 percent less energy than a conventional oven

Ceramic range tops use 15 percent more energy than conventional tops

Check door seal to see if heat is being wasted

Cover pots when heating water or food on the stove

Heat only the amount of water you need

Turn off electric heat early—burners will stay hot several minutes

Broiler oven settings uses much more energy per minute than bake does

Thawing frozen foods in oven wastes energy; thaw at room temperature

Washing Machine

Ninety percent of its energy use is for hot water; wash in cold water as often as possible

Use the low water level for small loads to save energy

If your washer has a water/suds saver feature, use it

Use short cycle to save energy whenever possible

Clean filter often

Dryer

Electric ignition on gas dryer saves energy

Automatic temperature-sensing saves energy

Use no-heat dry cycle whenever possible

Vent the dryer to the outdoors in summer; vent electric dryer inside in winter

Don't use dryer for one or two pieces unless absolutely necessary

Use "solar dryer" (clothesline)

Dishwasher

Wash only full loads

Use shortest cycle that will do the job

Cancel drying cycle and open dishwasher door to dry dishes

Use booster heater if your dishwasher has one and keep water heater at 120°F

Don't use your dishwasher as a dish warmer

General Tips

A color TV uses about three times more energy than a black-and-white TV

The instant-on TV feature uses more energy than a regular switch

Turn on cold water rather than hot when using garbage disposal

Electric blankets (or better yet, down comforters) save energy because you can keep your bedroom cooler at night

Lighting Economy

Most of us waste a large percentage of the energy used for lighting. Admittedly, the cost of lighting can't compare with that for heating, hot water, or major appliances. If you think you'd like to save that large percentage of your light bill, the process is mostly just common sense:

Don't use brighter lights than you need

Turn out lights not in use

One 100-watt bulb gives more light than four 25-watters

Fluorescent bulbs use much less electricity than incandescent ones

Remodeling Opportunities for Saving Energy

As this chapter shows, there's a lot you can do even in existing houses to save energy. And if you plan to remodel, the possibilities really blossom. Let's start right up front, at the entrance to your house. New, prehung steel doors are available with R-14 foam insulation. Special wood "warm doors" provide similar insulation. Both are far better than the R-3 or so a typical 1½-inch or 2-inch door gives. Also, consider a glassed-in entryway. This not only adds a bit of elegance but also can function as a sunspace to collect solar heat on clear days.

Table 1-13
Approximate Energy Use of Appliances

Appliance	Kwh/Mo	Appliance	Kwh/Mo
Air conditioner (window)	116	Heater (portable)	15–30
Air conditioning (central)	620	Hi-fi	9
Attic fan	90	House heating	1,000–2,500
Blanket	15	Radio	5–10
Blender	1	Range	100–150
Broiler	8.5	Refrigerator	25–30
Clock	1.4	frost-free	180
Clothes dryer	92	Sewing machine	1–2
Clothes washer	9	TV, black & white	24
Coffee maker	12	TV, color	42
Dishwasher	36	Toaster	5
Freezer	30–125	Typewriter	5
Frying pan	15	Vacuum cleaner	6
Garbage disposal	1	Water heater	200–300
Hair dryer	1–6		

If you plan to give your house an exterior face-lift, think about upgrading the insulation in the walls at the same time. You can use foam slabs, fiberglass batts, or some kind of blown-in insulation. While such an upgrading might not be economical if done alone, insulation costs can be shared with the renovating work you plan anyhow. If you're adding a room, this may be the ideal time to incorporate passive solar design. See Chapter 3 for more on passive heating and cooling. At the very least, maximize south windows and keep north windows small. Remember that awnings, overhangs, and trellises provide shade for summer cooling.

Starting from Scratch— The Superinsulated House

If you have the very good fortune to be planning not just a remodel but a new house, you have the golden opportunity to take energy conservation to its maximum. You're in the envious position of being able to take full advantage of the tremendous knowledge gained in the last 15 years—knowledge that means you can now build a house that you can almost literally heat with a match and cool with an ice cube! Such superinsulated houses match the performance of the best of the new solar houses.

The typical superinsulated house merits the name, with R-values greatly in excess of even a very well insulated house. Windows are kept small, particularly on the north side; some designs have no windows on this wall. In winter, lights and appliances may be sufficient to keep the house comfortable. In summer, the insulation keeps heat out, and little or no air conditioning is necessary.

Earlier in this chapter a caution was given against adding too much insulation. However, some new home designers carry insulation R-values far beyond normal recommendations. This superinsulation approach is in fact considered superior to passive design by some designers. The result is a house with very effective insulation, extensive vapor barriers, exceedingly tight construction, and very small windows—most of them south-facing. Hundreds of such houses are built each year and reward their owners with remarkably low heating and cooling bills. The more airtight a house is, the more important it is to use air-to-air heat exchangers, to rid the house of pollutants and odors and to keep sufficient fresh air inside for comfort and safety. This equipment uses electric power but partially pays its cost by recovering heat (or cool air) from the exhaust air.

Appealing as the superinsulation concept is, common sense suggests that it will be of most value in very cold climates and may be too much of a good thing in moderate climates. The extra initial cost may not be justifiable where heat loads are moderate. Aesthetics and personal tastes for outside view and natural lighting also enter the decision process.

For More Information on Conservation

The Complete Energy-Saving Home Improvement Guide, James W. Morrison, 1978: Arco Publishing Co., 219 Park Ave. South, New York, NY 10003

This large-format paperback is a compilation of useful government reports on retrofitting homes for energy conservation; even how to heat your house during an emergency.

The Energy Saver's Handbook, Massachusetts Audubon Society, 1982: Rodale Press, 33 E. Minor St., Emmaus, PA 18049

Focuses on conservation for town and city people.

The Fuel Savers: A Kit of Solar Ideas for Existing Homes, Dan Scully et al., 1978: Brick House Publishing Co., 34 Essex St., Andover, MA 01810

Small but useful primer. This little book covers not only conservation but construction and solar energy tips as well. Loaded with helpful drawings.

Movable Insulation, William K. Langdon, 1980: Rodale Press

The best book I've found on window insulation; it's complete, accurate, understandable, and easy to read.

The Tighter House, Charlie Wing, 1981: Rodale Press

A very useful collection of do-it-yourself conservation projects from *Rodale's New Shelter* magazine for weather stripping, insulation, heating, cooling, and plumbing.

Suppliers

Prices and addresses given are as of this writing; check with the supplier before sending money.

Night Insulation Hardware and Kits

Aerius Design Group, Inc. (RFD 1, Box 394B, Kingston, NY 12401) manufactures a Slip-In-Panel. It's a lightweight, highly insulating board with soft, springy edges for a good seal, made to be trimmed and decorated by user. The R-value is 8. It costs less than $2 per square foot.

Creative Energy Products (1406 Williamson St., Madison, WI 53703) makes Window Warmers. They are do-it-yourself insulated roman shades of decorator fabric, vapor barrier, Thinsulate insulation, and a drapery lining. Their R-value is 4. The instruction book costs $3.50 and packaged kits of material (except decorator fabric) cost $1.50 to $2 per square foot.

Dirt Road Co. (RD 1, Box 122, Waitsfield, VT 05673) offers Comfort Shades made of four layers of air-spaced reflective opaque film. The shades are cut to size at the factory and shipped to the customer with all hardware. Their R-value including two panes of glass is 7.7. Contact the company for the latest price.

Rockland Industries, Inc. (Box 17293, Baltimore, MD 21203) makes Roc-Ion Warm Window roman shades. Sewing instructions and kits are available for 20-, 40-, and 80-square-foot windows. The shades' R-value with one pane of glass is $7.69. They cost $3.50 to $4 per square foot.

Zomeworks Corporation (Box 25805, Albuquerque, NM 87125) makes Magnetic Nightwall Clips to hold night insulation panels in place. The 3-inch clips cost 43 cents each; the 6-inch clips cost 61 cents each. The company also makes Nightwall Spring Fingers for holding heavy or large night insulation in place. Spring Fingers cost 90 cents each.

Solar Domestic Water Heating

We all spend a sizable fraction of our home energy dollar on domestic water heating. On average it's about 20 percent, and if you have a houseful of active children, the percentage may be larger. That's why you should think seriously about solar water heating right after you take care of cutting energy losses. You need domestic hot water all the time—not just when you need space heating, or space cooling, but 365 days a year. Your solar water heater isn't a part-time investment; it will work for you every day.

Economics

A good commercial solar water-heating system completely installed and properly warranted can cost $4,000 or more (although many good systems are much cheaper). You can buy a new gas or electric water heater for a small fraction of that, so why opt for solar heating? The reason is life-cycle costing. If you're not familiar with the term, add it to your vocabulary because much of this book is based on this realistic method of considering not just initial cost but cost over the lifetime of renewable-energy equipment.

Chapter 1 discussed the advantages of buying a refrigerator (and other appliances) with a high EER or Energy Efficiency Rating even though the purchase price is more than Brand X. The economic reason for spending extra money up front is that within a given number of years you'll save more than that in lowered energy costs. That same logic applies to renewable-energy equipment; in the long run, spending the extra money should save you more energy dollars than it costs.

Generalizations are misleading, and you've got to look at your

What You'll Learn in This Chapter

- How it can pay you in the long run to spend 10 times as much or more for a solar water heater than a conventional one (page 57)

- How federal and state tax credits can save you up to 75 percent of the cost of your solar water heater (page 60)

- The differences among batch, thermosiphon, and active water heaters (page 62)

- That while a solar collector should ideally face true south, if it makes for neater and easier mounting, you can orient it 20 degrees or more east or west of true south with not too much penalty (page 67)

- That most solar water heaters can and do freeze unless they are properly protected (page 76)

- How much hot water you need a day, and how big a collector you should install (page 78)

- That you might be able to build a solar water heater large enough to supply all your hot water, but that you shouldn't (page 78)

- Why solar pool heating is the toughest water-heating job (page 82)

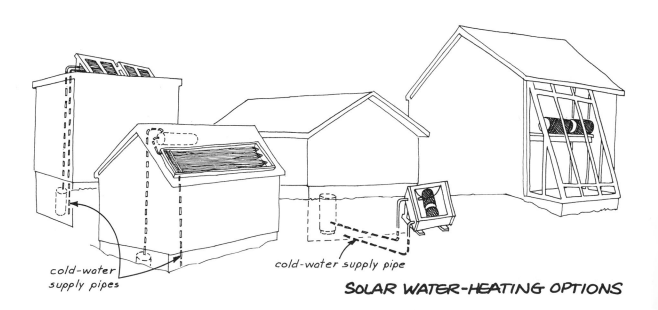

cold-water
supply pipes

cold-water supply pipe

SOLAR WATER-HEATING OPTIONS

own situation before opting for solar hot water. For instance, if you live in Washington or Montana and pay only 2 cents a kilowatt-hour for electricity, or if you have no shower-loving youngsters draining the tank a couple of times a day, your hot-water bill may be so small that a $4,000 solar water heater won't pay for itself in a reasonable time. But if your present water-heating bill is the more typical $25 or $35 a month, and going up, read on.

A hot-water bill of $35 a month will add up to $4,200 over the next 10 years—without any increases in energy costs. And indications are that energy cost increases will continue to run ahead of general inflation.

Assuming that conventional water heating is going to cost 15 percent more each year, today's $420 annual bill for hot water will rise to about $1,700 a year in 10 years. That sounds impossible, but get out your calculator and see. The total amount paid for hot water in the next 10 years (if this year's cost is $420) will be $8,526. Saving 75 percent of that (which is about what a good solar water heater can save you) will pay for a $4,000 solar water heater, plus interest and maintenance, and leave a little profit. In the 11th year, you'd save 75 percent of $1,955, or $1,466. And we haven't yet considered the tax credits for renewable energy.

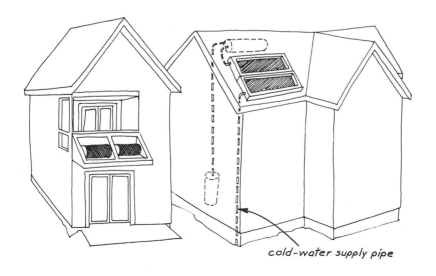

cold-water supply pipe

Here are six options for solar water-heating systems, from left to right: collectors on the roof and storage in the basement; a thermosiphon system with a SolaRoll collector; a freestanding batch heater; a batch heater in a sunspace; a batch heater built into a door overhang; and a thermosiphon system with a conventional collector.

Can You Save Water-Heating Costs by Going Solar?

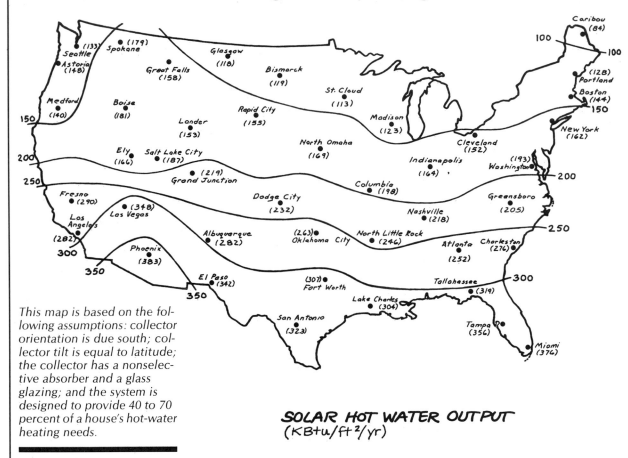

This map is based on the following assumptions: collector orientation is due south; collector tilt is equal to latitude; the collector has a nonselective absorber and a glass glazing; and the system is designed to provide 40 to 70 percent of a house's hot-water heating needs.

SOLAR HOT WATER OUTPUT
(KBtu/ft²/yr)

Tax Credits

Back in 1973, the energy crisis jolted Congress so badly that our lawmakers granted federal tax credits to those who install energy-saving equipment. As a result, Uncle Sam will pay 40 percent of the cost of your solar water heater (and other renewable-energy expenses as well).

Be sure you understand that the federal credit is a *bottom-line deduction,* subtracted not from your taxable income but from the tax that you owe. Simply put, the tax credit pays 40 percent of the $4,000 you spent for a solar water heater, bringing its cost down to only $2,400—which means it'll pay for itself much sooner and start paying you a return on your investment. Many states also grant tax credits in addition to the federal credits for renewable-energy installa-

The contour lines on this map will help you to make an initial estimate of the energy savings created by a solar water-heating system. The numbers appearing with the lines and those next to the city names all stand for solar energy contributions in the thousands of Btus per square foot of solar collector (KBtu/ft²/yr). Thus, an entry of 150 equals 150,000 Btu/ft²/yr. If you are planning a system with a certain amount of collector area, you can make a quick check on how much hot water it will actually deliver. You can also use the map as a planning tool to see how much collector area you need to supply your water-heating load. The match is simple: square feet of collector × KBtu/ft²/yr = annual output.

Say, for example, that you are planning to install 80 feet of collector in Dodge City, Kansas, which bears the number 232. You can expect to gain around 18,560,000 Btu (80 × 232,000) in solar Btus every year. From Chapter 1 you've already figured out your energy cost per million Btu (MBtu), so you can multiply that cost times the solar contribution (18.56 MBtu in the example) to get annual dollar savings. If you are heating your water electrically at a cost of $15/MBtu, the solar system will save you $278.40 every year (if electricity prices don't rise, which they do. The more they rise, the more you save).

tions. See Appendix C for a detailed explanation of how both of these credits work.

Solar Water-Heater Primer

At its simplest, a solar water-heater system consists of:

Collectors

Hot-water storage tank(s) (which, in some systems, is actually the collector as well)

Hot-water distribution system; in other words, plumbing

Controls for operating the system

Because it isn't practical to design a system for 100 percent solar hot water, most systems have a backup or booster water heater, which is generally a conventional gas or electric water heater.

Types of Solar Water-Heater Systems

There are two basic solar water-heating systems: passive and active. The passive type is simpler because it has no pumps or other moving parts; water moves either by gravity or by convection (hot water, like hot air, rises). The two most common passive designs are batch, or bread-box, heaters and thermosiphon heaters. Active systems have pumps that move water around, generally from a conventional flat plate collector to a storage tank.

Batch Water Heaters

These solar water heaters are so simple that the collector and storage tank are combined. The tank is painted flat black and absorbs heat, which is conducted into the water. The tank and reflectors (if used) are placed in an insulating box with a glazed cover to keep heat inside. If the box has an insulating lid that can be closed over it at night or during bad weather, the unit is called a breadbox heater.

Rodale's New Shelter magazine (May/June 1981) found that its batch heater is an excellent performer, with good durability and good freeze protection in all but very severe climates. Tested against four other designs, the batch heater showed a return on investment of 19 percent, tax-free. The heater was built for about $630 and saves about $300 a year.

Thermosiphon Water Heaters

In a thermosiphon water heater, the solar collector is separate from the storage tank, but water flows between them by convection and doesn't require pumps or other hardware. However, a thermosiphon system does require the storage tank to be 1½ feet or more higher than the solar collector for thermosiphon water flow. Thermosiphon systems generally do not provide any freeze protection and must be shut down and drained in winter. The collector in a thermosiphon heater is usually a regular flat plate collector.

Active Solar Water Heaters

Most commercial solar water heaters are active systems. These use collectors separate from their storage tanks and require pumps and controls to deliver hot water from collector to tank. In all but the warmest climates, freeze protection is a must; the two most typical freeze-protection systems are explained in detail on page 76 in this chapter.

Solar Basics

There's a tremendous amount of solar energy out there, but it won't do you much good if your climate is cloudy most of the time, or if your house is shaded by trees or buildings. If you do get enough sunshine, you must next be sure that solar collectors are properly sized for your hot-water needs and climatic region, and oriented so that they'll receive as much sun as possible.

A simple batch collector is little more than one or more water tanks that are painted a dark color and are set inside a glazed box. The inside of the box is often insulated and can be lined with reflective material, such as aluminum foil, to bounce as much heat as possible into the tank(s).

New Solar Water-Heater Types

Phase-Change Solar Water Heaters

Phase-change solar collectors use a Freon refrigerant instead of water or antifreeze in the pipes connecting the solar collector with the storage tank. When the sun shines on the collector, the liquid refrigerant boils into vapor (this is the phase change) and rises from the collector to the storage tank heat exchanger. Here it condenses to liquid, as it gives up heat to the water, and flows back to the collector to be heated again. Passive phase-change heaters, or thermo-siphoning systems, have tanks mounted above the collector, which somewhat limits their installation possibilities. Active phase-change systems don't have this limitation.

Proponents claim that the phase-change principle collects about 25 percent more solar heat than a conventional flat plate collector. Positive freeze protection is also provided, as the freezing point of Freon is −130°F! Basically the system works something like a refrigerator. It's hermetically sealed, and maintenance is claimed to be minor. Expert welding and other skills are required, so phase-change water heaters are not suitable projects for most do-it-yourselfers.

Heat Pump Water Heaters

Heat pump water heaters use electricity but produce more heat than conventional electric water heaters for the kilowatt-hours of power used. They can't compete with natural gas water heaters at present gas prices, but give heat pump water heaters a look; they cost less than a solar water-heater system of the same capacity. (For more

For starters, look at the map on page 60. This will give you a pretty good idea of how much solar energy per square foot of collector area your locale receives in comparison to the rest of the country. The higher the number (represented in thousands of Btus), the more solar radiation or insolation your location gets.

Collector Orientation

Ideally, your solar collector should face true south, not compass south. Check the map for compass variation where you live and use it to correct the compass reading. If the variation is 10 degrees east (as it is in central Texas, for example) *subtract* 10 degrees from the

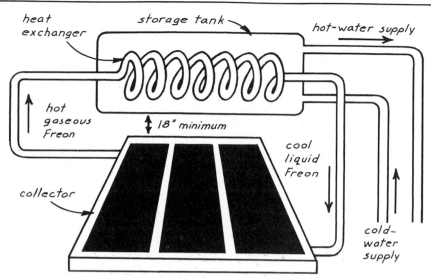

PASSIVE PHASE-CHANGE WATER-HEATING SYSTEM

A phase change is an alteration in form, such as from a liquid to a gas, or from a solid to a liquid. A material undergoing a phase change absorbs and releases more heat than a substance that simply heats up or cools down without physically changing. So it makes sense that a phase-change system such as this one would be more efficient than a regular solar heating system. But phase-change systems are controversial because they are expensive and difficult to install, and, in some cases, haven't worked well enough to justify the extra expense and bother.

The phase-change material in this passive system is Freon. The Freon starts out as a liquid, enters the collector, where it is heated to a boil, then becomes a gas. The hot gas is carried by convection through a heat exchanger, where it condenses back to a liquid, releasing solar heat to the water in the storage tank. Gravity returns the reliquified Freon back to the base of the collector.

information on heat pumps, turn back to the box "Heat Pumps", on pages 24 and 25 in Chapter 1.)

Photovoltaic-Powered Solar Water Heater

In this clever system the sun not only heats water but provides the electric power for circulating it. In addition to providing power for the pump, the photovoltaic panel also replaces conventional timers or sensors that turn the pump on and off: when the sun shines on the photovoltaic cells in the morning, electricity from them starts the pump; when the sun goes down, pumping stops.

reading on your compass. If you live in eastern Pennsylvania, the variation is 10 degrees west, and you must *add* that amount to whatever you read on your compass to find true rather than magnetic direction.

There are a number of shortcuts for locating true south. The sun itself is due south at solar noon. You should be able to get the exact time of solar noon for a given date from your local weatherman or an astronomy student. Or you can check the newspaper for the times of sunrise and sunset, calculate the length of the day, and divide by two. Drive a stake in the ground, and at solar noon the shadow will be exactly true north and south. South is toward the sun, unless you

A solar collector needn't stick up on your house in a highly conspicuous spot and be an eyesore. It can be located practically anwhere on or around your house, provided the glazing faces close to true south.

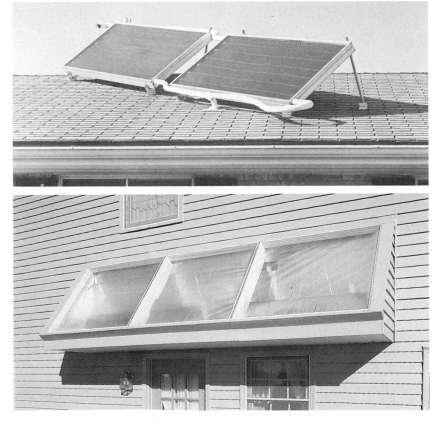

live in the Southern Hemisphere. For best performance, follow this RULE OF THUMB for collector orientation:

Orient your collector within 20 degrees east or west
of true south.

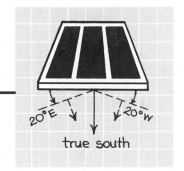

The orientation of a solar collector isn't nearly as critical as you might think; your collector, for example, can be off 20 degrees and sacrifice only 6 percent of the solar radiation. Check table 2-1 for the effect of collector orientation on collection efficiency.

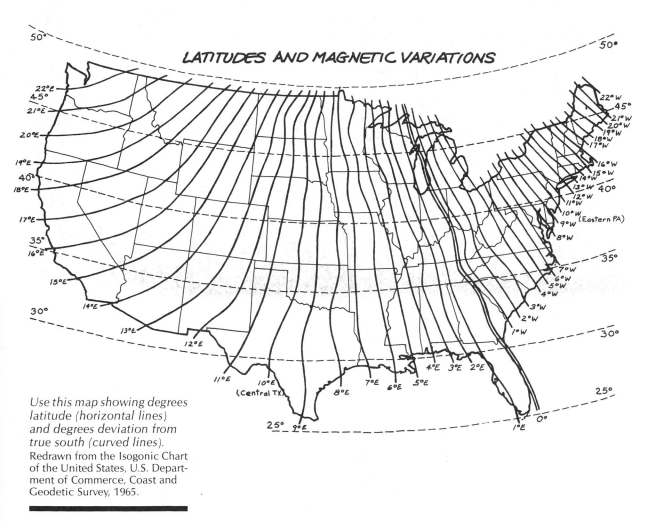

LATITUDES AND MAGNETIC VARIATIONS

Use this map showing degrees latitude (horizontal lines) and degrees deviation from true south (curved lines). Redrawn from the Isogonic Chart of the United States, U.S. Department of Commerce, Coast and Geodetic Survey, 1965.

Table 2-1

**Solar Radiation Available at
Various Collector Orientations**

Degrees East or West of True South	Radiation Received (%)
0	100
10	98
20	94
30	86
45	71
60	50
70	34
80	17
90	0

A PROTRACTOR FOR MEASURING SUN ANGLES

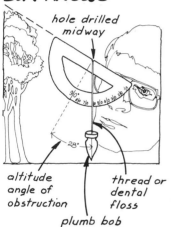

This simple-to-make angle finder will help you determine the altitude angle of nearby buildings, trees, and other objects that may block your access to sunlight. Take a plastic protractor, drill a hole exactly midway along the straight edge and tie thread or dental floss in the hole. Then tie washers or a plumb bob to the end of the thread for weight. Point the straight edge at the top of the object. The string will cross the degree mark that shows the protractor's tilt. That tilt is the object's altitude.

Solar Access

Some states have sun rights laws to guarantee that solar collectors keep working. California, New Mexico, and Colorado pioneered sun rights, and some other states have followed their lead. Check with your state energy office to see if such protection is available to you.

Don't think only of the present situation. Even if your house is bathed in sunlight from dusk to dawn this year, are you sure that won't change sometime in the future because of nearby new construction or maturing shade trees? A little shade is all right, but if a big tree comes between your south roof and the sun for a good part of the day, you'll have to trim the tree or forgo solar water heating. Winter is the season of most concern because that's when the sun is lowest in the sky and most likely to be blocked. If even part of the south-facing portion of your roof is constantly in the sun, you will probably have no solar access problem for a solar water heater. If you're not sure, better check for anything that could block your planned collector site from the sun.

With orientation stakes marking due south and the angle over which you'd need access to the sun, you can check on possible obstructions. To do this most accurately, use sun angle charts that tell how high the sun is in all directions at different times of the year. Charts for latitudes of 28, 32, 36, 40, 44, and 48 degrees are included here. They all look pretty much alike, so be sure to use the one closest to your latitude.

We'll use 40°N latitude and go through an example. The map of magnetic variations on page 67 will give you your approximate latitude. Orient the 40-degree sun angle chart properly by matching up

(Continued on page 72)

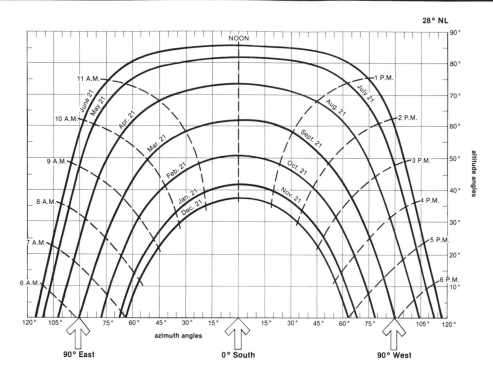

28° NL

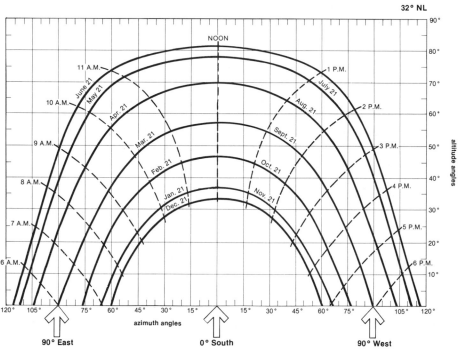

32° NL

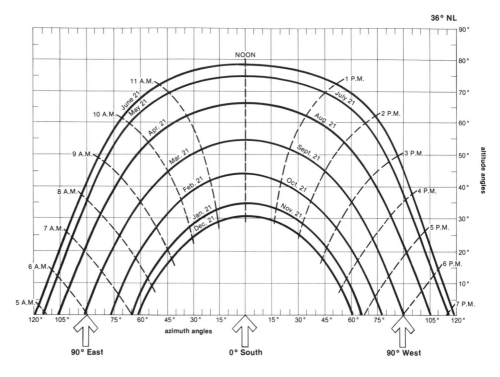

36° NL

NOON

11 A.M.
June 21
May 21
10 A.M.
Apr. 21
9 A.M.
Mar. 21
8 A.M.
Feb. 21
7 A.M.
Jan. 21
Dec. 21
6 A.M.
5 A.M.

1 P.M.
July 21
2 P.M.
Aug. 21
Sept. 21
3 P.M.
Oct. 21
4 P.M.
Nov. 21
5 P.M.
6 P.M.
7 P.M.

altitude angles

90°
80°
70°
60°
50°
40°
30°
20°
10°

120° 105° 75° 60° 45° 30° 15° 15° 30° 45° 60° 75° 105° 120°

azimuth angles

90° East   0° South   90° West

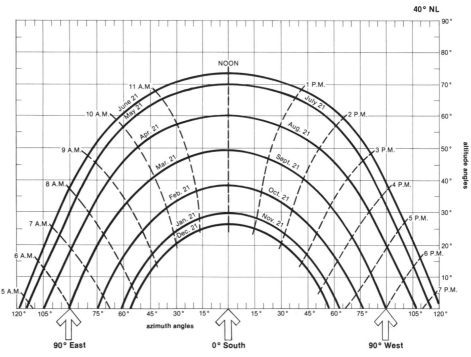

40° NL

NOON

11 A.M.
June 21
May 21
10 A.M.
Apr. 21
9 A.M.
Mar. 21
8 A.M.
Feb. 21
7 A.M.
Jan. 21
Dec. 21
6 A.M.
5 A.M.

1 P.M.
July 21
2 P.M.
Aug. 21
Sept. 21
3 P.M.
Oct. 21
4 P.M.
Nov. 21
5 P.M.
6 P.M.
7 P.M.

altitude angles

90°
80°
70°
60°
50°
40°
30°
20°
10°

120° 105° 75° 60° 45° 30° 15° 15° 30° 45° 60° 75° 105° 120°

azimuth angles

90° East   0° South   90° West

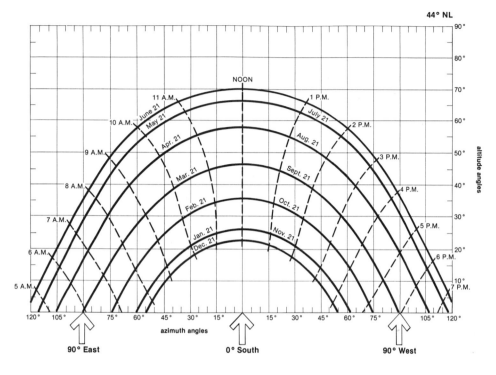

44° NL

NOON

11 A.M.
June 21
May 21
10 A.M.
Apr. 21
9 A.M.
Mar. 21
8 A.M.
Feb. 21
7 A.M.
Jan. 21
6 A.M.
Dec. 21
5 A.M.

1 P.M.
July 21
2 P.M.
Aug. 21
3 P.M.
Sept. 21
4 P.M.
Oct. 21
5 P.M.
Nov. 21
6 P.M.
7 P.M.

altitude angles

90° · 80° · 70° · 60° · 50° · 40° · 30° · 20° · 10°

azimuth angles

120° 105° 90° 75° 60° 45° 30° 15° | 15° 30° 45° 60° 75° 90° 105° 120°

90° East · 0° South · 90° West

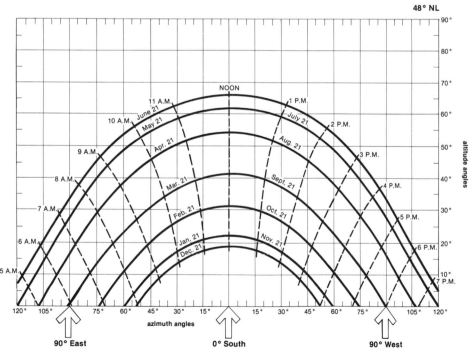

48° NL

NOON

11 A.M.
June 21
May 21
10 A.M.
Apr. 21
9 A.M.
Mar. 21
8 A.M.
Feb. 21
7 A.M.
Jan. 21
6 A.M.
Dec. 21
5 A.M.

1 P.M.
July 21
2 P.M.
Aug. 21
3 P.M.
Sept. 21
4 P.M.
Oct. 21
5 P.M.
Nov. 21
6 P.M.
7 P.M.

altitude angles

90° · 80° · 70° · 60° · 50° · 40° · 30° · 20° · 10°

azimuth angles

120° 105° 90° 75° 60° 45° 30° 15° | 15° 30° 45° 60° 75° 90° 105° 120°

90° East · 0° South · 90° West

Solar Site Analysis Using the Fist Method

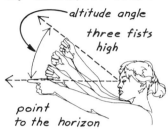

FIST METHOD OF ORIENTATION

altitude angle

three fists high

point to the horizon

The Energy Task Force of New York City, in *No Heat, No Rent* (New York: Energy Task Force, 1977), describes the fist method for determining whether buildings or other objects to the south of a roof would shade it and therefore render it useless as a site for a solar collector. This quick and easy method, which requires only two steady arms, plots the position of the sun on December 21 when it is at its lowest point in the sky.

First, you must determine the latitude of your site by consulting a map such as the one on page 67. Next, stand at the edge of the roof where your collector will be located and face true south. (Take

Table 2-2
Solar Sighting Angles

Your Latitude	True South	30° East and West of True South
28°N	4½ fists	3 fists
32°N	3½ fists	2½ fists
36°N	3 fists	2¼ fists
40°N	2½ fists	2 fists
44°N	2¼ fists	1½ fists
48°N	2 fists	1½ fists

SOURCE: Energy Task Force, *No Heat, No Rent: An Urban Solar & Energy Conservation Manual* (New York: Energy Task Force, 1977).

the south line on the chart with true south at your site as you have established it using one of the methods described earlier. Then follow the arrow out along the line to "noon." Notice the words "altitude angle" on the right side of the chart. On December 21, when the sun is lowest in the sky, its altitude (or angle above horizontal) at noon is about 26 degrees. At 8 A.M. and again at 4 P.M., the sun's altitude is about 5 degrees. Now you know the angles below which trees or other obstructions will not shade the solar collector.

With this information, and using the protractor and plumb bob method shown in the drawing (or using a commercial angle finder), any obstructions that will shade the collector can be found. Anything higher than the angles given on the charts for various directions will block the sun. Remember to check sun angles from wherever you plan to mount your solar collector. On the roof it's less likely to be blocked by trees or other obstructions than if mounted on the ground.

If you'd rather not bother with the precise sun angle method described above and want just a quick and dirty orientation method

care not to fall off!) Make a fist with your left hand and extend the arm, pointing to the horizon with your index finger (see the drawing). Your line of vision and your index finger will be parallel (horizontal). Make a fist with the right hand and place it on top of the pointed finger of the left hand. Alternate fist on fist the number of times indicated for true south according to your latitude in the table (sighting over the top fist indicates where the sun will be at noon). For example, in Knoxville, Tennessee, which has a latitude of 36 degrees, any object above three fists will shade the collector and any object below will be no problem. Obstructions just slightly above the last fist will not significantly affect the collector's performance throughout the year. Each fist, by the way, equals a 10-degree change in the altitude angle.

To determine if there will be obstructions 30 degrees east or west of the collector site (which shows the sun's position at about 9 A.M. and 3 P.M.), use the same method you used for true south, but consult the table again because the number of fists will be different. Let's use Knoxville again as an example. At 30 degrees east or west of south, any object above 2¼ fists will shade the collector.

In *The Complete Handbook of Solar Air Heating Systems* (Emmaus, Pa.: Rodale Press, 1984), Steve Kornher says that if a collector is to be ground-mounted, to use the fist method you must lie on the ground or stand 10 to 12 feet south of the area where the bottom of the collector will be located.

that works quite well, consider one developed by the Energy Task Force in New York City. It requires only two steady arms and will tell you sun angles for December 21 when the sun is lowest.

Collector Tilt

Ideally, a solar collector should be mounted so that the sun's rays strike it perpendicularly. But since the sun moves, you have to pick a compromise tilt that will deliver the most Btus during the day. Here's a collector tilt RULE OF THUMB:

> Tilt collector toward the south as many degrees from
> horizontal as your latitude.

If you mount the collector on your roof, which is generally the best place for it, you'll need to know the roof pitch to check for tilt. To find the pitch of your roof, use the table. Measure how many

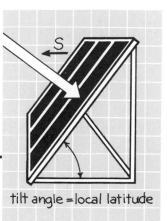

tilt angle = local latitude

HOW TO MEASURE ROOF RISE

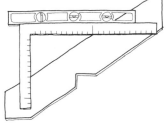

In order to determine the tilt of your roof collector, you must know the pitch of your roof. Run a level 12 inches out from the roof. Make sure it is perfectly horizontal. Now measure from the outside bottom corner of the level down to the roof. That is the rise. Use the table to find your roof's pitch angle. Now you can calculate the angle at which your collector should be mounted by taking the roof pitch angle and subtracting it from the latitude at which you live. For example, if you live at 40 degrees latitude and your roof pitch is 27 degrees, the collector's mounting frame should be built at a 13-degree angle (40 degrees minus 27 degrees).

Table 2-3
Finding Roof Pitch

Rise (in)	1	2	3	4	5	6	7	8	9	10	11	12	13	14	15	16	17	18	19	20
Pitch (deg)	5	10	14	18	23	27	30	34	37	40	42	45	47	49	51	53	55	56	58	60

inches of rise there are in 12 inches measured horizontally. Then check table 2-3 for the degrees of tilt. Now you know how close your roof comes to the ideal collector tilt.

Appearance Counts

Many solar water-heater installations fall far short of looking as good as they could—and should. We've talked about orientation and tilt, and the fact is that exact alignment isn't all that important. Yet many rule-bound installers insist that collectors be mounted at exactly the correct tilt. Besides sometimes looking bad, collectors tilted at a degree greater than the roof angle cost more to mount and aren't as well protected as those close to the roof. A collector mounted flat on the roof is not only unobtrusive, but also safe from the wind and won't let snow pack under it.

System Components

As pointed out earlier, a typical solar water-heater system includes:

Collectors   Distribution system
Storage tank(s)   Controls

Let's look at these components in more detail.

Collectors

A flat plate collector can heat either air or water. Air collectors need heat exchangers to transfer solar heat from air to water, and this results in a more complex system and also lowers the overall heat-collecting efficiency. For these reasons, most solar water-heater systems use collectors that heat water directly.

Flat plate collectors generally are made of metal, because metal absorbs heat and conducts it quickly to the air or water to be heated. Metals used include iron and steel (stainless steel is especially good since it won't rust), copper, and aluminum. Some collectors are made of plastic, but these generally are for low-temperature applications such as swimming pool heating, covered later in this chapter.

Solar collector absorber plates are usually dull black to absorb solar heat better. Ordinary flat black paint does a pretty good job, but there are much better absorbers, called selective surfaces. The best commercial solar collectors generally have selective coatings on the absorber.

As the drawing shows, the absorber plate also has passageways for the water it heats. Cold water generally comes in at the bottom of the collector, is warmed by the sun, and flows out at the top of the collector. Nearly all flat plate solar collectors are enclosed in an insulated box to minimize heat loss.

Glass or plastic glazing lets in solar energy but prevents much of the heat from coming back out through the glass. Good plastic glazings are made for solar collectors, but glass lasts for many decades without serious deterioration and is generally a wise choice for solar collectors.

Storage Tanks

Most solar water-heater systems have storage tanks in the house, basement, or garage. These tanks hold the solar-heated water and are called preheaters because the solar-heated water in them must be boosted up to a higher temperature by conventional water heaters. Often the original water-heater tank is used as the booster heater and is connected to the solar preheater tank. Hot water flows from solar collector to preheater, and then to the main tank for use. If only one tank is used, it should be large enough to hold all the hot water the collector can produce in a day.

Distribution System

Unless you're using a thermosiphon design, as described on page 62, you'll need an electric pump to circulate water between tank and collector. A very small pump is sufficient for residential applications; as little as 1/100th horsepower has been used. Some method of turning the pump on and off at the proper times must also be provided.

Insulation of the pipes and storage tank or tanks is critical; you don't want to lose hot water you've worked so hard to get. Here are two RULES OF THUMB:

At least R-4 insulation should be used on pipes running
from the collector to the tank.

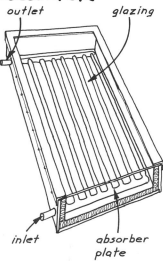

SOLAR FLAT PLATE COLLECTOR

outlet   glazing

inlet   absorber plate

A flat plate collector consists of a shallow insulated box, metal absorber plate with tubing or troughs, and glass or plastic glazing. Sunlight passes through the glazing and is absorbed by the collector plate. Cold water enters the system through an inlet near the bottom of the collector, flows through tubing bonded to the collector plate or along troughs in the plate, is heated, and flows out at the top of the collector through insulated pipes to an insulated storage tank somewhere in the house.

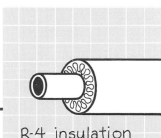

R-4 insulation

COMMON FREEZE-PROTECTION SYSTEMS

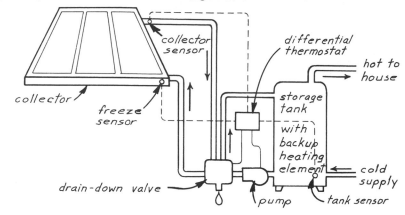

In the **drain-down system,** the key component for freeze protection is an electric drain-down valve. When the freeze sensor indicates that the temperature is approaching 32°F, the differential thermostat cuts electric power to the drain-down valve, which prevents water from flowing to the collector. At the same time, a small drain port in the valve opens, and the water in the collector drains down and through the drain-down valve to the house drain. The drain-down valve also protects the system during a power failure and if the collector overheats.

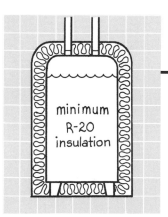

minimum R-20 insulation

At least R-20 insulation should be used around the storage tank if it's located outside or in an unheated place.

Controls

It would be wasteful to have a solar water-heater pump running at night, or when the sun isn't shining brightly enough to heat water. To prevent this, the pump can be manually controlled, with you doing the controlling. The control can be a simple clock-timer arrangement or a sophisticated, temperature-sensing control system with sensors on the collector that turn the pump on only when it's needed. Other sensors may turn on freeze protection when the temperature drops.

Freeze Protection

Freeze protection is recommended for practically all solar water heaters, except for batch heaters whose large quantity of water in

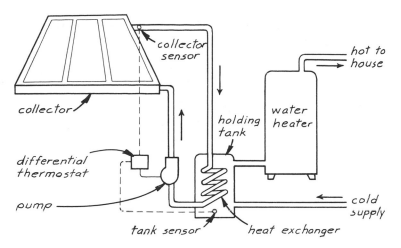

Unlike the drain-down system, a **drain-back system** does not operate by
an electric valve. A typical drain-back system has a solar plumbing loop,
which is a closed loop connecting the holding tank to the collector, and a
second plumbing loop, which connects the water heater to the heat exchanger
in the holding tank. The system is protected from freezing by gravity — if the
collector and pipes are properly positioned. When the collector sensor and
tank sensor indicate that the temperatures in the collector and holding tank
are the same or when the collector sensor indicates that there is insufficient
solar gain to be of benefit in heating water, the pump shuts off and the water
in the collector and pipes (the solar loop) drains by gravity to the holding tank.

the collector/storage tank prevents freeze-ups. A drain-back system,
using an antifreeze solution such as propylene glycol or ethylene glycol
between the collector and water tank, offers freeze protection; so does
a drain-down system that is designed with controls that drain the
collector of water when subfreezing temperatures occur.

Overheating

Freezing isn't a problem in summer, but overheating can be, particularly if the collector isn't used and is allowed to stagnate under
the hot sun. Without water circulation or air venting, collector temperatures can reach 400°F — in rare cases, some with wooden frames
have burned. If you put your solar water heater out of service in
summer, it must be vented, covered, or otherwise prevented from
overheating. This is an important point to discuss with a dealer when
buying a solar water heater.

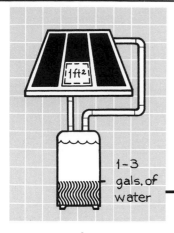

Collector Sizing

Now let's find out how large a collector (or collectors) you'll need. Fortunately, this calculation has already been done for you. A rough collector-sizing RULE OF THUMB is:

A good solar collector will make between 1 and 3 gallons of hot water a day (depending on season and geographical location) for each square foot of collector area.

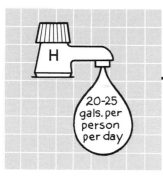

And a hot-water-use RULE OF THUMB says:

The typical American family uses 20 to 25 gallons of hot water a day per person.

Let's put the two rules together. Since a four-person family uses 80 to 100 gallons of hot water a day, 30 to 90 square feet of solar collector should keep them happily in hot water. More accurate estimates require adjustments for service water temperature, climate, and available sunshine. And that's where the Solar Index comes in.

The Solar Index

The Solar Index program, developed several years ago by Colorado State University (CSU) for the U.S. Department of Energy, takes care of those regional sizing details. Using the SOLCOST computer program, CSU specialists calculated the collector area necessary in various cities to provide 75 percent of the energy needed on average to heat 80 gallons of water a day (enough for three or four people) from groundwater temperature to 120°F. Collector sizes for representative U.S. cities are given in table 2-4.

Notice that the collectors weren't sized to provide 100 percent of the hot-water needs. It's not cost-effective to do this, because several consecutive days of bad weather would necessitate a huge solar collector and a very large storage tank to guarantee hot water all the time. So the 75 percent fraction is a good compromise. This means that you'll need a backup source for hot water when the weather is bad—or when you need extra hot water.

Table 2-4
Solar Index Collector Sizes for Family of Four

City	Ft²	City	Ft²	City	Ft²
Birmingham, Ala.	50	Louisville, Ky.	65	Fargo, N.Dak.	75
Flagstaff, Ariz.	40	New Orleans, La.	45	Cleveland, Ohio	80
Yuma, Ariz.	28	Bangor, Maine	75	Oklahoma City, Okla.	55
Little Rock, Ark.	60	Baltimore, Md.	80	Portland, Oreg.	90
Los Angeles, Calif.	35	Boston, Mass.	70	Philadelphia, Pa.	65
Oakland, Calif.	50	Detroit, Mich.	80	Providence, R.I.	70
Santa Maria, Calif.	45	Minneapolis, Minn.	80	Charleston, S.C.	73
Denver, Colo.	55	Jackson, Miss.	55	Sioux Falls, S.Dak.	65
Bridgeport, Conn.	75	Kansas City, Mo.	60	Knoxville, Tenn.	60
Washington, D.C.	55	Great Falls, Mont.	80	Austin, Tex.	45
Wilmington, Del.	70	Omaha, Nebr.	60	Dallas/Ft. Worth, Tex.	45
Miami, Fl.	40	Las Vegas, Nev.	40	Salt Lake City, Utah	50
Atlanta, Ga.	50	Concord, N.H.	80	Burlington, Vt.	90
Boise, Idaho	60	Trenton, N.J.	65	Norfolk, Va.	60
Chicago, Ill.	80	Albuquerque, N.Mex.	40	Bellingham, Wash.	100
Indianapolis, Ind.	70	Buffalo, N.Y.	75	Charleston, W.Va.	73
Dubuque, Iowa	70	New York, N.Y.	62	Madison, Wis.	80
Dodge City, Kans.	50	Charlotte, N.C.	60	Cheyenne, Wyo.	60

The Solar Index gives a fair estimate of how much collector area you'll need no matter where you live. The complete Solar Index tables (which are not included here) also state that Phoenix and a number of other cities don't need freeze protection. But unless you live in Hawaii or another tropical place, better play it safe and include freeze protection wherever you install a solar water heater.

Buying a Solar Water Heater

There are many kinds of solar collectors and solar water-heater systems. Most of them are good; some may not be so good. As with any major purchase, spend enough to get a good system—and then be sure to get all that you pay for. A collector with a good selective coating on the absorber, and two panes of glass rather than one, will heat more water than the same size collector with black paint and only one thickness of glass—especially in a very cold climate.

As important as the collector components themselves is the quality of the installation. In the early days of solar, many problems resulted from poor installation and lack of understanding of how the system worked. But a lot has been learned in the last 30 years, and standards have been established for components and installations as well. Check with your state energy office for applicable standards, and make sure that the dealer you select is in compliance. Some states insist on this before granting the tax credits mentioned earlier.

Remember the collector sizing rules of thumb, and don't let some fast-talking salesman sell you a 100-square-foot collector array for just you and your spouse. Don't let an even faster talker convince you that his superspecial model produces enough hot water for a family of six with a collector only 20 square feet in area and priced accordingly. As in any business, you generally get what you pay for. Remember these RULES OF THUMB:

It's wise to deal only with properly licensed contractors and, if possible, to buy close to where you live. Insist that the contractor take out a building permit. Get an owner's manual and study it before you buy. Talk turkey about price and get as good a bargain as you can, but be wary of bargain sales and special deals that seem too good. Be just as choosy about financing, too.

Find out about certification and the warranty: be sure that you get them in writing. Ask for names and addresses of people who have bought the product, and check their opinions. Most people with solar equipment are proud of it and enjoy talking about it and maybe even showing it off. If they've had problems with it, they may tell you that, too! It's better to learn from their mistakes than repeat them.

A good, knowledgeable installer may be the most important "part" of your solar water-heating system. The next best part is probably the owner's manual.

Building Your Own Solar Water Heater

If you'd rather build than buy, you have three options: building from scratch, building from a kit, or building in a workshop class. The most challenging is to pick a set of plans and start from scratch. An easier way is to buy a kit of parts and assemble them according to instructions. This costs more than the first option but should save time and perhaps result in a better solar system.

Best of all, you may find a workshop class where you can build a commercial-grade collector under expert supervision and also get information on installing and operating it. These classes are offered at some universities and colleges; some designers and solar energy associations also sponsor them. Building your own solar water-heater system under workshop supervision is easier than the first two options and may not be any more expensive.

Comparing the Collectors

Rodale's New Shelter magazine (May/June 1981) made exhaustive tests on five home-built solar water heaters and ranked them in a number of categories. Table 2-5 gives performance of all the designs. Notice that the least cost-effective type still showed a tax-free return on investment (ROI) of 9 percent, indicating it should repay its cost in about 11 years. Again, don't forget that every solar water heater will perform better if conservation measures are used.

The systems took from six to nine days to build, with the phase-change heater taking the least amount of time and the batch heater taking longest. The phase-change design was also ranked most difficult; it should not be attempted by the average do-it-yourselfer.

Building a Batch Solar Water Heater

New Shelter's winning system was a batch heater; its cost was low and it was relatively easy to build. It's a simple design because it combines the solar collector with the storage tank. With no circulating flow of water to and from the tank, the heater heats a batch at a time. These heaters are generally used as preheaters, with the water temperature then boosted by another tank heated with gas or electricity.

The Solar Hot-Water Life-Style

You have a potential money saver in a solar water heater, but if you aren't careful, the backup heating system can wipe out those

Table 2-5
Ranking Five Home-Built Solar Water Heaters

Type	Cost/Ft² ($)	Total Cost ($)	ROI* (%)
Batch	15	630	19
Drain-back	17	1,506	17
Drain-down	21	1,053	13
Thermosiphon	13	674	13
Phase-change	26	3,500	9

SOURCE: "And The Winner Is . . . . ," *Rodale's New Shelter* (May/June 1981).

*Return on investment

savings. You must learn to work with what you've got—and what you've got is solar; it differs greatly from conventional water heating. Solar energy is available only when the sun shines. When it's cloudy, there are fewer Btus going into the storage tank; at night there are none.

Suppose that your collector sits there in the bright sun for six hours heating gallons and gallons of water. The tank is full of hot water by 2 P.M., but since you don't use any until 5 P.M., a lot of solar energy goes to waste.

After dinner, you wash a week's worth of clothes and follow with a dishwasher run. Next morning, the showers are skimpy—or would have been, except that your backup heater turned on automatically. That makes the gas or electric meter go around so fast that at the end of the month you wonder why solar energy didn't provide the savings you expected.

You'll probably use less backup heat by rescheduling tasks to use some hot water earlier in the day. Wash clothes and maybe the breakfast dishes, if they are a sizable load, after lunch. Then the solar tank can take on more solar heat so there'll be hot water for evening baths without an expensive assist from the backup system.

Solar Heat for Swimming Pools

For a long time solar water heaters were used mostly for swimming pools. Only in the last couple of years have domestic hot water and space heating taken the larger share of the market. The popularity of solar pool heaters is understandable because of the high cost of heating pools with natural gas or other fuels.

The Toughest Solar Heating Job

Most people greatly underestimate the requirements for heating pools. It's a big order for two reasons: first, a very large volume of water must be heated. Second, pool water loses heat very rapidly during cold weather.

The same heat losses that plague your home have an even greater effect on your pool. The water *radiating* its heat to the cooler sky accounts for about 10 percent of the total heat loss. Wind blowing over the water carries away 20 percent, and 5 percent is *conducted* from the bottom of the pool into the ground. There's another villain in the form of *evaporation*. When water turns to vapor, it absorbs great quantities of heat. Evaporative cooling can keep you cool in

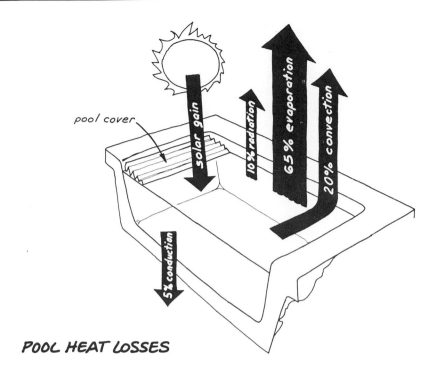

pool cover

solar gain

10% radiation

65% evaporation

20% convection

5% conduction

POOL HEAT LOSSES

About 95 percent of a pool's heat losses occur at the water surface. Evaporation accounts for the biggest heat loss, followed by convection (water taken up with the wind) and radiation. Only a small amount of heat is lost to the ground by conduction.

summer, but it lowers pool temperature fast in winter, causing 65 percent of the heat loss!

The situation is much like that of a very leaky house. So before thinking about adding a solar pool heater, you should do everything you can to cut down on heat losses from the pool.

The sun must provide about 56,000 Btu a day for heating domestic hot water for a typical family of four, but that's only a drop in the bucket toward heating a swimming pool. In San Francisco in May, the average minimum air temperature is about 50°F and the high temperature is about 70°F. Water in a swimming pool typically stabilizes at about the average of those two temperatures, or 60°F.

If you know how many gallons of water your pool holds, the arithmetic is simple. If not, you'll have to calculate the volume. As an example, let's use a rectangular pool 16 feet wide and 30 feet long. It's 3 feet deep at one end and 8 feet deep at the other. Average the depth at 3 + 8 ÷ 2, or 5½ feet. Then 16 × 30 × 5½ = 2,640 cubic feet (and about 20,000 gallons) of water. Since a cubic foot of water weighs 62.4 pounds, the pool contains 164,736 pounds of water and needs a lot of Btus to heat it for winter swimming.

Multiplying pounds by 20 (the degree difference between what the water temperature is, 60°F, and what you'd like it to be, 80°F) gives 3,294,720 Btu, considerably more than the 56,000 Btu needed to heat domestic water for a family of four! It's obvious that a pool can't be heated in a day with solar energy; the task requires not just good solar-heater design but many days of heat storage as well.

Fortunately, we don't have to heat pool water to the 140°F or so for home use; about 80°F is generally plenty for swimming. A second advantage is that the swimming pool itself is a sizable solar collector.

Pool Covers

Just as the glazing on a solar collector stops some of its heat from escaping, a plastic pool cover cuts heat losses from a swimming pool. Evaporation and wind cooling losses are greatly reduced. This simple and inexpensive fix can add several degrees to a pool's temperature, so a cover is the easiest and cheapest way to heat a pool.

The best pool covers are made from double-wall bubble plastic. This has the added advantage of being an insulating blanket to keep in even more heat; the old trapped-air trick. Many versions of this kind of pool cover are available commercially; the best have ultra-violet inhibitors in the plastic for long life. Because of the inconvenience of rolling up the cover, some owners cover the pool with lengthwise strips of bubble plastic and remove just one strip so they can swim laps.

Solar Pool Heaters

Solar collectors for pools are usually low-temperature plastic collectors costing much less than conventional solar collectors. Their operation, care, and maintenance are similar to those for solar water heaters, but they probably won't last nearly as long. Here's a RULE OF THUMB:

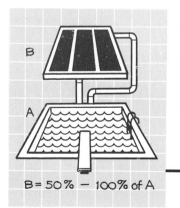

B = 50% — 100% of A

To be effective, collector area should be 50 to 100 percent of pool area, depending on environmental conditions.

For example, in Palm Springs, California, a 480-square-foot pool can be heated with about 240 square feet of collector. In Denver or Des Moines, however, swimming might be chilly even with 480 square feet. At $5 a square foot, such collectors will cost from $1,200 to

It takes a lot of sunshine to make a solar pool-heating system such as this one work at its peak efficiency. The pool cover, partially unrolled here, helps prevent heat loss at night and on cloudy days.

$2,400. Remember that a pool cover helps collectors work much better and can reduce the collector area required.

Solar Heating for Spas and Hot Tubs

Spas and hot tubs can be heated with solar energy, too. Both require a much higher temperature than a swimming pool, but less collector area because of the smaller volume of water to be heated. A commercial pool heater should be able to supply the approximately 110°F needed. Freeze protection is required if you use the spa or hot tub in winter, expecially if conventional metal solar collectors are installed.

Don't expect a solar pool heater to match the fast recovery of a large gas heater. Also, save as much heat as you can by using an insulating cover over the tub. For a RULE OF THUMB:

A hot tub 6 feet in diameter and 4 feet deep requires about 150 square feet of solar collector.

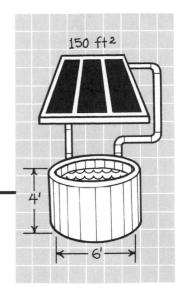

150 ft²

4'

6'

For More Information on Solar Water Heating

The Homeowner's Handbook of Solar Water Heating Systems, Bill
  Keisling, 1983: Rodale Press, 33 E. Minor St., Emmaus, PA 18049

> A state-of-the-art planning and insulation guide for do-it-yourselfers
> and those wishing to buy systems.

Passive Solar Water Heaters, Daniel K. Rief, 1983: Brick House Publishing Co., 34 Essex St., Andover, MA 01810

> A comprehensive how-to book for designing and building batch
> solar water heaters. Covers construction, plumbing, siting, and
> performance.

Solarizing Your Present Home, Joe Carter, ed., 1981: Rodale Press

> This book has an excellent section on water heating with solar
> energy. Covers theory, design, construction, and operation, plus
> several do-it-yourself projects.

Suppliers

Prices and addresses given are as of this writing; check with the
supplier before sending money.

Solar Water Heater Plans and Kits

ALR Energy Products (RD 1, Box 415, Belvidere, NJ 07823) offers a
flat plate solar water heater kit. The heater is 3½ feet by 6½ feet
and uses EPDM (rubberlike) tubing instead of copper to carry the water.
The kit includes an aluminum frame, fiberglass glazing, absorber plate,
and insulation. It costs $300. The company also offers a solar pool
heater kit. It has a 4-foot-by-8-foot collector panel, uses an aluminum
absorber plate, and EPDM tubing to carry water. It costs $140. The
header system (there is only one needed regardless of the number
of panels) costs $50.

Integral Design Solar Water Heating (4708 Raley Blvd., Sacramento,
CA 95838) offers plans for building a 40-gallon batch heater. The total
materials cost ranges from $350 to $450. The plans cost $5.50.

Our Solar System, Inc. (6704 Stuart Ave., Richmond, VA 23226) offers a drain-down solar water heater kit. It includes a 50-page instruction book, two 3-foot-by-7-foot black chrome flat plate collectors, and all solar components except the water tank. It costs $1,049. The company also conducts collector workshops. Write for details.

Rodale's New Shelter Magazine Project File (Box 155, Emmaus, PA 18049) offers detailed plans for a solar batch heater that can be built for less than $400 (including a federal tax credit). The plans cost $3.95. Ask for Plans Package number 112.

Solar Living, Inc. (Box 12, Netcong, NJ 07857) offers a drain-down solar water heater kit. It includes instructions; two 4-foot-by-8-foot collector kits (copper absorber plate, plastic glazing, aluminum frame, and Thermax insulation); pump control unit with valves, sensors, and thermometers; air vent/relief valve package; and one 82-gallon water tank. It costs $1,560.

A closed-loop solar water heater kit is also available. It is the same as the kit described above except that an antifreeze solution is used in the pipes from tank to collector. It costs $1,580.

Solar Usage Now, Inc. (Box 306, Bascom, OH 44809) offers the Arizona State University/DOE solar water-heater kit used in many university workshop classes. It includes parts for two flat plate collectors, pumps, heat exchanger, controls, gauges, valves, and plumbing fittings. The kit costs $965 plus shipping costs for distances more than 150 miles. The company also offers several other kits. Kit 3880 includes two finished 34-inch-by-76-inch SUN collectors, one 80-gallon insulated preheater tank, heat exchanger, control, and pump. It costs $1,250. Kit 3882 is the same as 3880, except it has three collectors. It costs $1,555. The Btu Bucket kit includes a 2-foot-by-8-foot collector box, aluminum absorber plate, copper tubing, and fiberglass glazing. Insulation is not included. This collector costs $100. The high-quality hydronic collector measures 34 inches by 77 inches and uses a copper absorber plate and waterway. The kit is sold without glazing or insulation. Standard glass patio doors (single-or double-glazed) are recommended. The kit costs $185.

Sunergy (7190 Fiske Rd., Clinton, WA 98236) offers plans and instructions for their Skyheat 30-gallon batch heater. The plans cost $20. A minikit of hard-to-find hardware and plumbing items costs $50, and the completed heater sells for $700.

Sunkeepers Thoughtful Solar Applications (1488 Sandbridge Rd., Virginia Beach, VA 23456) offers passive solar water-heater plans. Step-by-step plans for a batch-type preheater cost $6 a set.

Zomeworks Corporation (Box 25805, Albuquerque, NM 87125) offers plans for building two different batch solar water heaters without expensive tools. Write for current plan prices. The company also has the Big Fin absorber plate kit, which includes 8-inch-by-96-inch aluminum absorber plates for use in building your own unglazed solar water heaters. Big Fins are made for use in a greenhouse or sunspace and generally do not require freeze protection. They can also be used in thermosiphon or pumped solar water-heater systems. Write for current prices.

Passive Solar Heating and Cooling

The term "passive solar energy" was unheard of in the early days of "modern" alternative energy, back in the beginning of the 1970s; it wasn't part of the solar vocabulary then. But look how far we've come in that short time: many housing experts now believe that any new house that doesn't incorporate passive heating and cooling options is obsolete before the last nail is driven and the final coat of paint brushed on. And the owners of many existing houses have learned what an effective technology passive is.

Earth itself is the prime example of passive heating and cooling. Without the sun, it would be a frigid ball of ice at a temperature near absolute zero, which is about −459°F. Fortunately, the sun does shine and global temperature averages a fairly comfortable 59°F.

Solar heat is stored in the ground, in water, and in the air to help keep us warm. In summer the wind (caused by uneven solar heating) and the evaporation of water help keep us cool. So the sun already provides most of our heating and cooling needs. We can add the relatively little more that's needed by consciously using basic passive techniques.

Passive heating and cooling is the controlled use of solar heat and environmental cooling through the natural processes of radiation, conduction, convection, and evaporation. In passive solar heating, there's no mechanical apparatus separate and distinct from the house's structure as there is in active solar (which is covered in Chapter 4).

In passive techniques, the house is the solar heating system: its windows and other glass areas are the collectors; its walls, floors, and other massive components store the collected heat. Movement of solar heat from collector areas to storage and living spaces is by means of natural air movement, sometimes guided through vents and ducts.

What You'll Learn in This Chapter

- After conservation measures, passive solar heating is most likely your best option (page 93)

- There are two major passive solar systems: direct gain, which is the simpler; and indirect gain, which is usually more complicated but also more flexible, especially for retrofits for existing houses (page 97)

- Your house needn't face true south to perform well with passive heating (page 101)

- When it's best to use glass and when it's best to use plastic glazing in passive retrofit projects (page 103)

- The importance of thermal mass to prevent large hot-to-cold temperature swings inside your house, and how different materials rate as thermal mass (page 105)

- Good passive solar add-ons to a south-facing exterior wall are a masonry storage wall (also called a Trombe wall) and a sunspace; water walls can be retrofitted inside a large, south-facing window wall (page 112)

- Thermosiphoning air panels (TAPs), like masonry storage walls, are good exterior add-ons. They're usually less expensive than masonry storage walls and easier to build. Window box heaters, smaller versions of TAPs, are cheaper still but provide only supplemental heat to a room (pages 112–113)

- Thermal mass-to-glass ratios and orientation guidelines for sunspaces or attached greenhouses are the same as those for direct-gain designs (page 114)

- Except in the warmest climates, you shouldn't design a new home for 100 percent passive solar heating; how to determine the maximum solar heating contribution you should aim for (page 122)

- A passive solar-heated house is easier to cool than a conventional house because it is so weathertight; special features you can build into a new home or add onto the one you've got to keep it naturally cool in summer (page 123)

If the distribution of heat is helped along with an electric fan or blower, the system is technically a hybrid: passive collection and storage, with an active distribution system. Good passive designs not only provide solar heat in winter but also incorporate passive cooling elements that shield the house from unwanted summer sun and provide good ventilation so heat that does collect inside can escape outdoors.

Why Passive Heating?

Most of this chapter describes the addition of passive features to your present house; it's a primer on supplementing what you have now with the most cost-effective solar energy applications there are. Information on new passive solar houses is given later in this chapter.

Don't even think about passive heating until you've made your existing house as efficient as you can. And this means going back to Chapter 1 and measuring your energy conservation applications against those spelled out there. Caulking and insulating aren't as glamorous as adding a sunspace or thermosiphon air panel, but conservation is still the cheapest way to save energy in an existing house —even cheaper than passive heating.

It would be foolish to install a passive solar retrofit at considerable expense and allow all that wonderful, free warm air you're collecting to escape through an uninsulated attic, leaky windows, or ill-fitting doors. Solar energy stops when the sun goes down; the more

Adding solar can enhance a home's appearance—and its living space—if it's well integrated into the house's design.

A Quick and Dirty Estimate of the Solar Heating Value of Your Passive System

100   50

Caribou
(39)

Spokane
(125)

Glasgow
(65)

50

Seattle–
Tacoma (107)

Bismarck
(55)

St. Cloud
(33)

Portland
(86)

Astoria
(116)

Boston
(106)

Medford
(139)

Boise
(128)

Rapid City
(101)

Madison
(81)

New York
(95)

Lander
(98)

North Omaha
(80)

Cleveland
(87)

100

Ely
(126)

Salt Lake City
(129)

Grand Junction
(124)

Indianapolis
(83)

Washington
(121)

Columbia
(114)

Greensboro
(156)

Cape
Hatteras
(170)

150

Dodge City
(142)

Las Vegas (297)

Fresno
(159)

Oklahoma City
(146)

Nashville
(113)

Atlanta
(109)

Charleston
(182)

200

Los
Angeles
(210)

Albuquerque
(189)

Phoenix
(282)

North Little Rock
(96)

250

El Paso
(216)

Fort Worth
(179)

Tallahassee
(165)

250

Lake
Charles
(148)

San Antonio
(158)

150

Tampa
(168)

200

This map is based on the following assumptions: collector orientation is due south; collector tilt is equal to latitude plus 15 degrees; the collector has a nonselective absorber and a glass glazing; and the system is designed to provide 40 to 60 percent of a house's heating needs.

SOLAR SPACE-HEATING OUTPUT
(KBtu/Ft²/yr)

heat your house leaks, the quicker your house cools off—or the sooner the backup heating system must come on.

If saving energy for heating and/or cooling is your aim, passive solar should get priority, after you've done your best with conservation. Here's why:

Reason 1: You've already got most of your passive system right there in your house: south-facing windows to collect heat, and floors and walls (especially if they're made of stone, tile, brick, cement, or concrete block) to store that heat

Like the map used for solar water heating on page 60 in Chapter 2, this contour map lets you make a quick estimate of the output of a passive space-heating system (such as a direct-gain window array or a Trombe wall). It works exactly like the solar water-heating map: the numbers next to contour lines and city names represent the solar output (heat actually delivered to the living space) in terms of thousands of Btus per square foot of glazed area per heating season. You can pick a number that is closest to your area and multiply it times the area of the passive glazing you're thinking of installing. You'll end up with a huge number, in the millions of Btus, so just divide the result by 1 million to get the result expressed in MBtus. That number can be multiplied by your present heating cost (in $/MBtu), which you figured out in the box "How Much Does Your Home Energy Really Cost?" on page 22. What you'll get for all your effort is the equivalent dollar value of the solar heat you'll get from your system. In other words:

$$\frac{\text{MBtu/ft}^2/\text{yr} \times \text{ft}^2 \text{ of collector area}}{1,000,000} \times \begin{array}{l} \text{present heating cost in} \\ \text{\$/MBtu} = \text{dollar value of} \\ \text{solar heating} \end{array}$$

If you live near the 100-KBtu contour line, the system you install will deliver about 100,000 Btu per square foot of glazed area per heating season. Thus, 200 square feet of direct-gain glazing will provide 20,000,000 Btu, or 20 MBtu, of space heating (200 × 100,000 = 20 MBtu). If you were heating with oil at a cost of $11/MBtu, you could expect a return of about $220 a year. Of course, before you actually make the investment and begin any major work like this, you should consult with a solar engineer or architect who can give you a more thorough and detailed analysis of solar performance for your area.

Reason 2: Costs will be low, and with a passive system, first cost is generally the last cost as well

Reason 3: Maintenance on windows, floors, and walls is much less than it is on pumps, fans, and electronic controls. Breakdowns just don't occur because there's nothing to break down

Reason 4: Because of the simplicity of passive techniques, you'll have a lot less clutter caused by mechanical contrivances, plumbing, and freezing and overheating protection—to say nothing of repairmen underfoot

Reason 5: The best has been saved for last: the quiet enjoyment of comfortable solar heating and the pleasure of using natural processes rather than mechanical devices. Letting the sun shine in takes on its finest meaning with passive solar heating

The Economics of Passive Heating

Solar energy is free, but putting it to use costs money. So, in spite of all the aesthetic benefits, most of us have to look beyond them at our budgets. Let's see if a passive retrofit will be cost-effective for you.

Add up utility bills, oil receipts, wood bills, or whatever else is necessary to get an accurate estimate of your annual heating cost. If it's very small, a passive retrofit may not be worth the bother. But if it's large enough to be a concern, look closely at passive.

Let's assume a total of $1,000 for present annual heating as a starting point in deciding if it will be cost-effective to upgrade an existing house with passive solar. Three years' worth of heating bills, compounded at 15 percent a year for fuel cost increases, will total $3,472. If passive heating saved 50 percent of your heating cost, a retrofit costing more than $1,750 would be a loss for you in the short term. Five years is more interesting; in that time you'll spend almost $7,000 for space heating, as shown here:

first three years	$3,472
fourth year	1,520
fifth year	1,750
Total	$6,742

If you save half those heating costs by spending $3,000 on passive features, you'll make a profit within five years. Your passive heating system will still be as good as new, so let's total up the heating bills for another five years:

first five years	$ 6,742
sixth year	2,011
seventh year	2,313
eighth year	2,660
ninth year	3,059
tenth year	3,518
Total	$20,303

Half of $20,303 is $10,151 saved for a passive retrofit outlay of only $3,000! Even if you spend $7,000 on the retrofit, the saving in 10 years will still be about $3,000. And we haven't yet considered the renewable-energy tax credits that can reduce the cost of your passive system. These federal and state tax credits are explained in Chapter 2 and Appendix C.

If you've already decided upon a passive retrofit, the box here will help you estimate its dollar value in solar heating.

Passive Heating Design Options

As was pointed out earlier, all houses that the sun shines on are already passively heated to some extent. Those that stay warm inside during the day primarily because of solar energy are often called sun-tempered. To be good passive solar houses, they must not only heat up during daytime, but store that heat for nighttime as well. To keep warm at night, a house must be well insulated and it must have sufficient thermal mass. The amount of storage depends on the house itself, the environmental conditions in the region where you live, and how warm you want to be.

There are two major design options for passive heating, direct gain and indirect gain. One of these, or some combination of both, can be used to retrofit your house.

Direct gain is the simpler approach, since it makes use of existing windows. Direct gain is nothing more than heating your house with the direct solar energy that comes into living areas through generally south-facing windows and other glazed areas. In addition to windows, this includes clerestories, skylights, glass doors, and sunspaces that are not separated from living areas by a wall or doors. If you have enough south glazing, your house can stay warm even on a very cold day.

With all its advantages of simplicity and heating effectiveness, however, direct gain has some disadvantages. These include sun glare through large areas of glass, fading of drapes and upholstery, overheating during the day, and excessive loss of heat through windows at night. Sufficient window area for passive heating may also seriously limit wall space on the south side of the house.

Good direct-gain design is achieved by understanding these problems and coping with them. For example, lightfast fabrics are available. There are also windows that contain film that blocks ultraviolet rays, so there is no fading of carpet, drapes, or furniture. Suf-

(Continued on page 100)

Because passive retrofits take many shapes, there's a good chance that at least one design will work well with your present house. Shown here are clerestories, solar greenhouses, and plenty of south-facing glazings that have been added to existing houses.

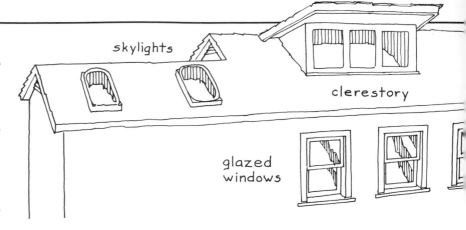

skylights

clerestory

glazed
windows

DIRECT-GAIN OPTIONS

Direct gain is the simplest way of letting the sun heat the interior of your home, while providing extra light at the same time. **Skylights** *are attractive alternatives, which can open up a room, even one with no exterior wall, to the outside. Operable skylights provide summer venting.* **Clerestories** *are built into the roof and are frequently used to bring light into back, north-facing rooms.* **Vertical wall arrangements** *include windows, sliding glass doors, and bay windows.* **Attached greenhouses** *that are open to the interior of your home offer extra living space and a place to grow flowers or food.*

Heat and light can be too much of a good thing if you don't design carefully. Good movable insulation (see Chapter 1) is effective at cutting down night heat loss, and shades and overhangs help to keep out unwanted summer sun. The more direct gain, the more thermal mass you need to prevent large interior temperature swings.

ficient thermal mass in floors and walls absorbs solar heat coming through the windows and keeps the house from getting too warm during the day. This stored heat then helps keep the house warm at night.

Indirect gain includes a variety of approaches in which direct sunlight doesn't enter the house. However, while indirect gain keeps direct sunlight out, it lets solar heat in. The heating is "indirect" because solar energy heats an existing wall or other thermal mass that stands between the sun and the interior of the house. This thermal mass then heats the living space by radiation, conduction, and convection.

Indirect-gain designs include thermal storage walls, thermosiphoning air panels (TAPs), and attached sunspaces or greenhouses that are separate from living areas. Each design offers several variations on the basic concept, making it easier to tailor indirect gain to your specific situation.

Indirect gain solves the problems of glare, fading, and useful wall space but has other problems of its own. No light comes through a solid wall, of course. And unless the exterior glazing is properly handled, it can be unattractive.

Passive Heating Basics

Now let's see if your existing house qualifies as a candidate for a passive retrofit. Two environmental conditions are necessary for passive heating: sufficient insolation, and access to that insolation. Insolation, or solar radiation, is covered in detail in Chapter 2. It varies with location and season; Phoenix gets a lot more than Buffalo does.

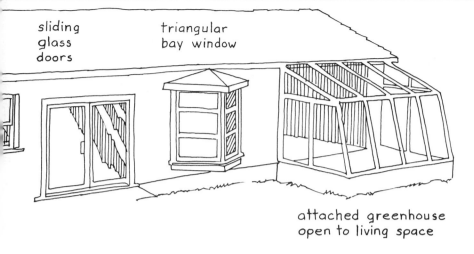

sliding glass doors

triangular bay window

attached greenhouse open to living space

The second requirement is that your house have access to insolation. A house with no sun shining on it isn't a candidate for passive heating, so don't try to retrofit for passive heating if the sun's rays don't reach you. Fortunately, most houses have some passive potential.

Appendix A gives other important values called degree-days. The need for heating and cooling is measured in heating degree-days and cooling degree-days, respectively. Degree-days are the number of days times the number of degrees below or above 65°F during those days. Some mild climates have so few heating degree-days that houses rarely or never need heating; Hawaii, for example. Bryce Canyon, Utah, on the other hand, has so few cooling degree-days that it almost never needs cooling. Knowing the amount of insolation available, and the amount of heating a house requires, makes it possible to design an effective system.

Collecting Solar Heat

We'll assume that you have a tight, well-insulated house on which the sun shines fairly regularly in winter. A handful of sunny days can't keep you warm all winter; be sure you're not fooling yourself. Some regions just don't get enough sunshine to make passive solar houses work, so carefully check the insolation data for your city.

If you've read Chapter 2, you know about orienting a solar collector for maximum heat gain. Should your house face true south, fine. But if that's not the case, don't give up. While a rectangular house with its long wall facing south does function very well for passive heating, one with a short wall facing south can also perform satisfactorily. So can a square house with one corner pointing south.

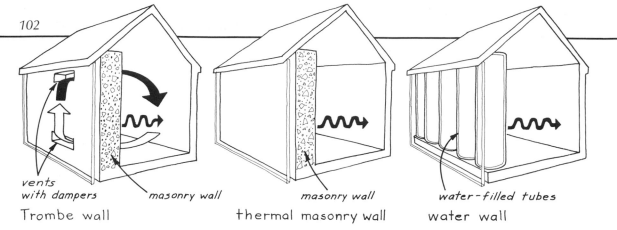

vents with dampers   masonry wall

Trombe wall

masonry wall

thermal masonry wall

water-filled tubes

water wall

INDIRECT-GAIN OPTIONS

Add triangular, L-shaped, round, and oval houses to the list and you'll begin to get the idea: just about any shape house may be a candidate for passive heating, although extreme designs will probably cost more to retrofit.

Letting the Sun Shine In

Glazing lets the sun shine in during the day and keeps some of its heat in at night, a sort of one-way valve for solar energy. The simplest glazing is an ordinary window. Windows that face south let in lots of direct solar energy; east and west windows let in direct sunlight in morning and afternoon. Windows facing north let in sunlight, too, but this diffuse kind isn't as strong as direct solar radiation. North windows represent a net heat loss.

Windows and glass doors and walls aren't the only ways to let in sunshine, of course. Sunspaces or attached greenhouses also let in heat and light. So do skylights, although historically they've been used mostly for light and not heat. They're difficult to insulate, and lots of heat can be lost through them at night.

Clerestories, vertical glass areas above an adjoining roof, are increasingly popular in new passive houses and can add much heat and light even to rooms on the north side. Like skylights, they are difficult to insulate, and because they're typically larger than skylights, the problem is compounded. A point to consider before going to the expense of adding clerestories to an existing house is whether or not you'll be able to reach them to clean the glass. Skylights generally aren't as much of a problem in this regard because the bottom side of the glazing won't pick up as much dust as vertical glass does.

A single pane of window glass has an R-value of less than 1, compared with up to R-19 for the wall it's set into. That's the reason

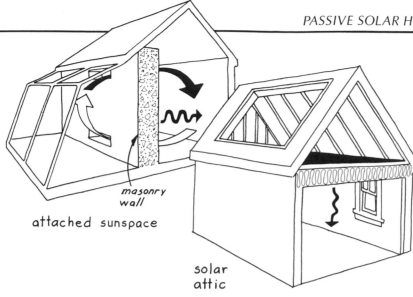

attached sunspace

masonry wall

solar attic

Indirect-gain systems have an advantage over direct gain in that they allow better control of light, heat, and drafts. Most indirect-gain systems incorporate the use of a wall or other thermal mass that absorbs solar heat and slowly releases it to living spaces. A **Trombe wall,** *which is a form of thermal wall, has vents that open in daytime for heat to pass through and close at night to prevent backdrafting. A* **thermal masonry wall** *heats a home entirely by radiation. It has no vents, so the heat takes some time to pass through the wall, and during the day, the warmth generated may be minimal.* **Water walls** *give off heat more rapidly than masonry walls, and if the tubes are translucent, they still let sunlight inside. An* **attached sunspace** *can heat up quickly, especially if it is not shaded properly or there is insufficient thermal mass to absorb excess heat. Vents aid convection and help prevent summer overheating. The* **solar attic** *absorbs heat at roof level, then fans move the heat through ducts down to living spaces.*

for double or triple glazing, which gives values approaching R-2 and R-3 respectively. However, don't use more than two glazings except in extremely cold environments. Besides the expense of extra panes, multiple glazings cut down the amount of solar energy coming through them.

Glass or Plastic?

Glass is attractive and, barring breakage, is also one of the longest-lasting building materials. High-quality glass transmits as much as 92 percent of the light striking it. Glass can be so transparent that it's almost invisible; translucent enough to offer a soft, diffused light instead of glare; opaque enough to provide privacy; or as reflective as a mirror.

Plastic glazings range from very thin, flexible films to rigid panels as thick as heavy plate glass. They are suitable for some passive retrofit projects. Plastics have a slightly higher R-value than glass and are warm to the touch.

The window frame is an important factor in heat loss. Metal is a far better conductor of heat than wood is, and much more heat escapes through ordinary metal sash to cold air outside. Wood sash offers better insulation, but it can rot if not caulked properly, is heavier, and costs more. You can eliminate the heat loss of metal sash by using the thermal-break kind, in which an insulating gasket separates the inner metal surface from that exposed to the outside air. Thermal-break sash has much better insulating properties.

Slate floors and water tubes,
if designed properly, provide
thermal storage, absorbing
heat during the day and
releasing it into living spaces
at night.

Table 3-1
Comparison of Various Heat-Storage Materials

Material	Relative Cost (the lower the number, the lower the cost)	Relative Storage Capacity (the lower the number, the greater the storage capacity)
PCMs	1	1
Masonry	2	5
Gravel	3	4
Earth	4	3
Water	5	2
The building itself	6	6

SOURCE: "A Handmade Invitation to a Star," *Harrowsmith* (November 1981).

Storing Solar Heat

Collecting heat is only the first step in making a passive house work. The second, equally important, step is storing that heat. Passive pioneers operated largely by guesswork, and many early passive houses were too successful at collecting heat and not successful enough at storing it. These bad designs baked their occupants by day and froze them at night.

On a sunny day, tens of thousands of Btus stream into a conventional house with lots of south-facing glass, temporarily turning it into an oven. But when the sun sets and the flow of solar heat stops, the house can quickly cool below the comfort level. The solution to this extreme temperature swing is to provide sufficient thermal mass in the form of masonry floor and/or walls, water tanks, or phase-change materials (PCMs) receiving direct sunlight. The National Research Council of Canada has ranked the various heat storage materials for relative cost and storage capacity. Note that PCMs are most expensive but store the most heat. Using the house itself is cheapest but least effective.

Whatever the thermal mass, a portion of the collected solar heat is stored in it. The stored heat is slowly released, and even after the sun goes down, it warms room air by conduction. Its radiant heat also warms the occupants directly, as the sun does when it shines on them. Radiant heat is pleasant heat and can make a room comfortable when the air temperature is considerably lower than that required for forced-air heat.

Masonry Heat Storage

Your existing house may already be capable of storing a large quantity of heat. A concrete slab floor, especially a dark one, collects

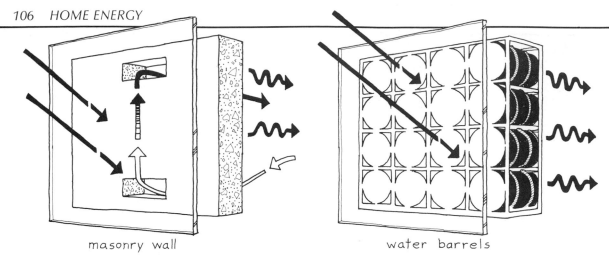

masonry wall   water barrels

*Everything in your house—
walls, floors, furniture—
stores a certain amount of
heat. This heat storage is
known as thermal mass. In
passive design, you want to
maximize this mass to prevent
sharp day-night temperature
swings, and to retain some of
the solar heat for nighttime
warmth. You do this by adding extra thermal mass. The
three best thermal-mass
mediums are masonry, water,
and the newest, phase-change
materials.*

*A **masonry wall** installed
behind glazing absorbs heat
and slowly radiates it to the
living area. Vents are often
cut into it to maintain a flow
of warm air.*

*   **Water** can absorb more
heat than masonry can, so
some people choose to use
water for thermal mass. The
water can be stored in barrels
or tubes. If the tubes are clear
or translucent, you can enjoy
sunlight as well as heat. But
rust and leaking can be
problems if the containers
aren't treated properly.*

THERMAL MASS OPTIONS

heat as sunshine strikes it. (But not through carpet or rugs!) Internal masonry walls or partitions reached by the sun also store heat. If you have plenty of heat-storing floors and walls, you're fortunate; if not, you must add some other kind of thermal mass.

Keep in mind dual-purpose applications as you consider adding needed thermal mass. Perhaps you've always wanted Spanish tile for its good looks but couldn't justify its cost and the labor involved in setting it. If the tile can also save on heating bills, you may feel you can afford it. An added heat-storage wall can also do double duty by forming a hallway along one side of a room. Brick is good flooring material and excellent thermal mass. Used as a wall, or as facing on an existing wall, it absorbs heat well, looks great, and never needs paint.

Water for Heat Storage

Water is a very good heat-storage material. Less volume is needed for water storage than for anything else except the phase-change materials covered later in this chapter. Water has its problems, though: it tends to corrode metal containers and then leak out of them. Many successful water-storage systems are sold commercially, however, including durable, lightweight, clear plastic tubes. This kind of water wall is very effective for storing heat and also for decorative purposes.

Phase-Change Heat Storage

In passive systems, heat is most often stored simply by raising the temperature of the storage material. This is called "sensible" heat storage. The melting and solidifying of PCMs stores and releases large quantities of heat through "latent" heat storage.

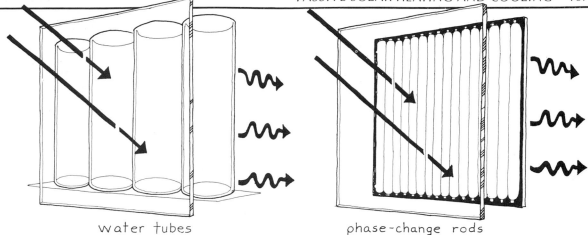

water tubes   phase-change rods

A typical PCM stores heat about as well as rock until it reaches the melting point. Then it may store 100 or more Btu per pound as it melts. After that, it reverts to sensible heat storage.

Melting ice absorbs large quantities of heat and is useful for cooling, but its 32°F phase-change temperature isn't in the right range for heating applications. Certain salts and other materials, however, have phase-change temperatures higher than that of water, thus storing and releasing heat at temperatures useful for passive heating.

Storing 1 million Btu at house-heating temperatures would require more than 900 cubic feet of masonry or rock or about 360 cubic feet of water. But a suitable PCM can store that much heat in a volume of only about 80 cubic feet. A container of PCM little more than 4 feet on a side thus holds as much heat as a rock bin nearly 10 feet on a side.

PCMs cost considerably more than rock or water, and there have also been some problems with their lifetime. However, PCMs are being improved and are worth your consideration. For example, the Solar Energy Research Institute is investigating materials called poly alcohols, or polyols, which undergo a phase change while remaining in the solid state. With their high heat capacity and very light weight, polyols may someday make excellent thermal mass. A heat-storage wall made of polyols would weigh only one-eighth as much as concrete with equivalent heat storage and be much smaller.

Phase-change materials *can be contained in plastic pipe or a latex skin. Although phase-change materials are not now widely used, they are promising because they store more heat in a much smaller space than either water or masonry.*

Preventing Heat Loss

After the collection and storage of solar heat, the third step in passive design is to retain as much of that heat as possible. If your

Table 3-2
Percentage of Heating Load Saved with Movable Insulation

R-Value of Movable Insulation in Addition to Window	Percentage Saved*	
	(with single glazing)	(with double glazing)
1	32	22
2	42	31
3	46	38
4	49	41
5	51	44
6	52	46
7	53	48
8	54	48
9	55	50
10	55	50
11	56	51
12	56	52

SOURCE: William K. Langdon, *Movable Insulation* (Emmaus, Pa.: Rodale Press, 1980).

*Based on temperature difference = 35°F with movable insulation over the window 14 hours per day

house is tight and well insulated, you've made a good start. But there's more to be done toward making it as effective as possible in retaining the heat stored in thermal mass. Especially since you may be adding sizable areas of glazing to collect more solar energy.

The basics of night insulation are covered in Chapter 1. Table 3-2 will give you an idea of the effectiveness of night insulation inside your glazing, showing approximate amounts of daily net heat gain for each 100 square feet of glazing in January at 40°N latitude. Notice that R-5 insulation more than doubles the heat gain for single glazing and adds about 60 percent over double glazing. A combination of double glazing and night insulation triples the heat gain of single glazing.

Designing Passive Heating Systems

In the few years that the science of passive heating has been developing, a great deal of design information has become available. The architect who is planning a new solar house draws on sophisticated computer programs and other design tools. The current bible is *The Passive Solar Design Handbook,* Vol. 3, prepared by Los Alamos Scientific Laboratories for the U.S. Department of Energy. Anyone

Table 3-3
Passive Collector Area Needed*

Coldest Average Outdoor Temperature Degrees F/Degree-Days per Month	South Glass as % of Floor Space to Be Heated	
	36°N Lat	48°N Lat
15/1,500	27	42 (NI)†
20/1,350	24	38 (NI)
25/1,200	21	33
30/1,050	19	29
35/900	16	25
40/750	13	21

SOURCE: Edward Mazria, *The Passive Solar Energy Book* (Emmaus, Pa.: Rodale Press, 1979).

*Table is for average house with about 10 Btu per square foot per degree F heat loss. If your house is very tight and well insulated, less glazing will be needed
†NI means night insulation is a must

serious about passive design should have a copy of this definitive handbook. However, it will tell most homeowners far more than they need to know to do retrofit passive projects. Anyone familiar with computer design techniques and wanting to use them as an aid to retrofitting a house should refer to Appendix C for a variety of available programs for hand-held calculators and home computers as well.

Design of passive heating systems for existing houses will depend primarily on what you have to work with. Using information from earlier sections of this book and the decision trees in the Introduction, including the RULES OF THUMB, you should be able to decide whether to use direct gain, indirect gain, or some combination of both.

For a quick estimate you can assume that:

> For each 10 square feet of floor area in a house, provide 1 to 4 square feet of south-facing glazing.

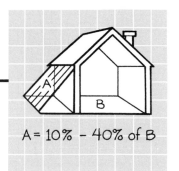

A = 10% – 40% of B

Table 3-3 will give you a more specific idea of how much south glass you'll need to adequately heat your house depending upon your latitude.

Let's look at a couple of examples. Sacramento, California, has an average temperature of 45°F and about 600 degree-days in January, so a passive house there will need between 11 and 17 percent as much south glazing as there is floor space to be heated. The note says that for northern latitudes the high percentage should be used. Sacramento is close to 40°N latitude, so 15 percent would be about the right amount of glass. Carrying out the design process, a 1,500-square-foot house would require 225 square feet of south-facing glass—not too difficult a project.

Madison, Wisconsin, has an average temperature of 17°F and about 1,500 degree-days in January, needing between 27 and 42 percent as much south glass as floor space. Since Madison is at 43°N, the maximum 42 percent ratio should be used. Notice, too, that the table calls for night insulation because of the very cold temperature. A 1,500-square-foot house here would need 630 square feet of south glass. This gives you an idea of the limitations of passive heating.

Direct-Gain Design

In a direct-gain passive heating system, all south-facing glass contributes solar heat. With an existing house, you must make the best of what you have unless you plan extensive and expensive modifications. Generally, the most cost-effective approach is to open up more window space and/or put in clerestories and skylights where possible. With a suitably shaped house, clerestories put heat and light into north-facing rooms.

The simplest passive design is the sun-tempered house, in which the sun shining through existing windows keeps the house comfortably warm during the day. For example, a conventional house in the Boston area with about 12 percent as much south-facing window area as floor space can get about 20 percent of its heat from the sun.

If you want more than daytime heating from the sun, your house must have ample thermal mass to store collected heat for nighttime comfort. As described earlier in this chapter, this mass can be existing masonry floors and walls, added masonry, water walls, or PCMs placed strategically to collect solar heat. Here are some heat-storage RULES OF THUMB:

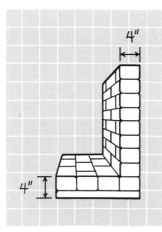

A thermal mass wall or floor should be 4 inches thick to retain sufficient heat for nighttime use. Concrete mass thicker than 6 inches is useless (except in Trombe walls).

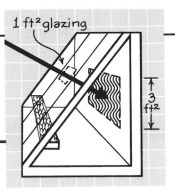

1 ft² glazing

3 ft²

For each square foot of south-facing glazing, a surface area of about 3 square feet of masonry thermal storage (walls and floors) is recommended for direct gain.

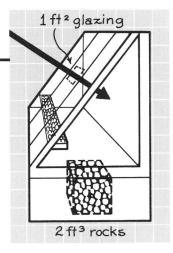

1 ft² glazing

For every square foot of south-facing glazing, use 2 cubic feet of rocks.

2 ft³ rocks

Heat is distributed through a house by the natural processes of convection, conduction, and radiation. Where necessary, small fans can be used to move heated air to rooms not adjacent to south glazing.

Indirect-Gain Design

Indirect-gain passive heating is the proper option where south glazing isn't available, or where direct sunlight isn't wanted in a living space. As noted earlier, there are a number of indirect-gain options:

Thermal storage walls
Thermosiphoning air panels (TAPs)
Sunspaces/greenhouses isolated from living areas

As with direct gain, heat is collected and stored in masonry, water, or PCMs.

Here are two thermal mass-sizing RULES OF THUMB:

Depending on climate, an interior thermal storage wall should be from 25 to 50 percent of house floor area for effective performance.

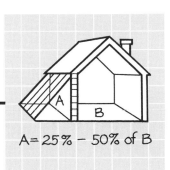

A = 25% − 50% of B

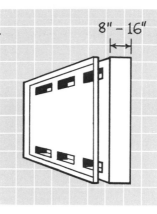

8" – 16"

A Trombe wall should be 8 to 16 inches thick.

Masonry Storage Walls

The most-used thermal storage walls are masonry, glazed on the outside and with an air space between glazing and wall. These are generally called Trombe walls. Glazing adds the greenhouse effect to a Trombe wall, which heats up and slowly transfers heat to the interior. To provide heat more quickly to the living space, vents are cut through the Trombe wall a few inches above the floor and a few inches below the ceiling, creating a circulating warm-air heater. Here's a Trombe wall vent-sizing RULE OF THUMB:

Make the total vent area 2 to 3 percent of the glazing area.

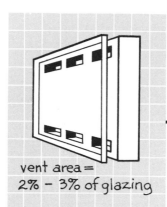

vent area = 2% – 3% of glazing

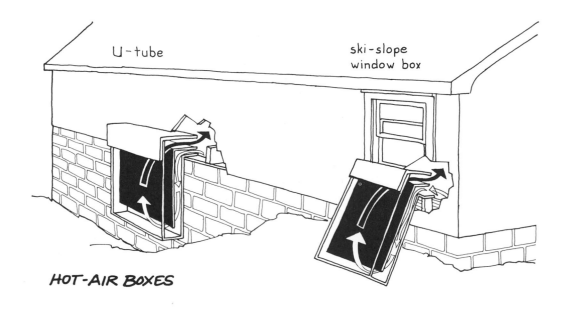

U-tube

ski-slope window box

HOT-AIR BOXES

Air in contact with the heated wall rises, displaced by cooler air from inside the house. Air flow can be controlled by adjustable louvers on the vents. When the room is warm enough, the louvers can be closed, letting the wall store more heat to warm the house at night. Adding night insulation, such as shutters or sliding panels, or insulation between glazing and thermal storage wall, keeps in more heat at night.

Water Walls

Early water walls consisted of discarded fuel drums painted black to better absorb solar heat. Now a variety of metal and plastic containers are available commercially. The water wall generally allows some direct solar heat to enter the room, and space around the water tubes allows convective air flow so there is no need for an inner wall with vents. However, night insulation should be provided to prevent

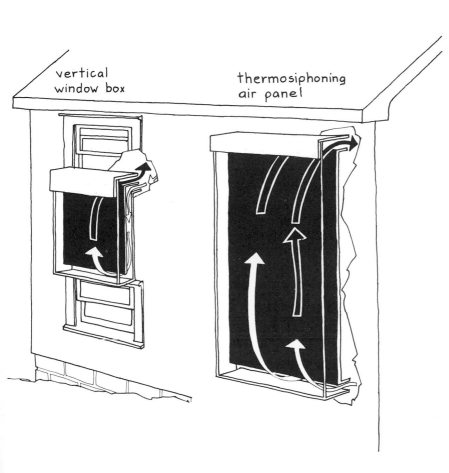

vertical window box

thermosiphoning air panel

Hot-air boxes are simple passive solar collectors. Although they will provide only a limited amount of heat to a room, they can be fairly easily retrofitted to a house. And you can increase heat gain by using a few per room, so long as you have enough south-facing windows. A **through-the-wall U-tube** *doesn't need a window for access to the living area; an exposed basement wall is an ideal mounting spot for these heaters. If you have a low first-floor window, a* **ski-slope window box** *is for you. Its tilt absorbs extra winter sun. A* **vertical window box** *is good for upstairs windows, and during the months it isn't needed you can remove it and store it. A* **thermosiphoning air panel** *(TAP) can adapt to the wall of almost any home and can span two stories, as shown here. Make sure you close the vents on a TAP at night to prevent reverse thermosiphoning (drawing warm house air into the cooling collector).*

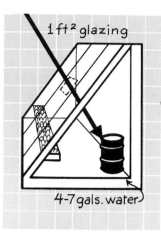

1 ft² glazing

4-7 gals. water

heat from going back out through the glass. Here's a water wall-sizing RULE OF THUMB:

For each square foot of south-facing glazing, provide from four to seven gallons of water in a water wall.

Thermosiphoning Air Panels (TAPs)

Thermosiphoning air panels don't incorporate any heat storage. Instead of thick masonry, water tubes, or PCM containers as in thermal storage walls, a TAP has a black sheet metal heat absorber behind the outer glazing. Vents cut into the house wall permit cooler air from inside to flow into the TAP, be heated by the absorber, and enter the house through the top vent. Many TAPs have small electric fans for better circulation of warmed air.

TAPs are very efficient solar heaters and transfer heat into the room quickly. Remember, however, that they operate only when the sun shines on them. When the sun goes down, be sure to close the vents so that heat from inside won't escape through them. Unless you can completely cover your south walls with TAPs, don't count on them to supply more than about 25 percent of needed space heating.

The Clearview Collector

A very effective variation of the thermosiphon air collector is the Clearview collector. Effectively a combination picture window and solar collector, it was developed by scientists at the University of Arizona's Environmental Research Laboratory in Tucson.

The Clearview collector is simply two panes of glass with a venetian blind between them. One side of the blind is painted a flat, dark color to absorb solar heat. Air between the glass panes is heated by the warm blinds and rises by convection as cooler air from the house replaces it. In hybrid versions of the system, a small fan moves warm air to more distant rooms.

Sunspaces/Greenhouses

The attached sunspace not only heats a house but also serves as a pleasant sun-room. This dual function makes it the most popular and most visible direct-gain passive design. Sunspaces—also called attached solar greenhouses, sun-rooms, atriums, and even walk-in solar collectors, are far more varied in design than thermal storage walls. In fact, about the only place you'll find two just alike is in a

This Clearview collector is under construction on a new house. At the right are the jambs that support the glass panels. To the left is the masonry wall, which stores the collected heat.

passive solar subdivision. Depending on the orientation of the house, they may be glassed-in front porches, central atriums, or sun-rooms added to the back of the house or to one side or the other.

To gather most heat, the south wall of the sunspace should be perpendicular to the rays of the sun at noon. Since you can't adjust the glazing to keep it perpendicular to the sun, you must select a tilt for the glazing and live with it all year. Here are three glazing-tilt RULES OF THUMB:

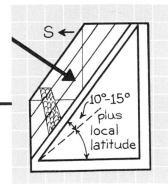

Tilt the south wall of the sunspace to local latitude plus 10 to 15 degrees.

(For example, if you live at 35°N latitude [Fort Smith, Arkansas; Winslow, Arizona; and Bakersfield, California, for example], your sun-space's south wall should be tilted up 45 to 50 degrees from the horizontal.)

(Continued on page 118)

*Attached sunspaces are as
varied as personal tastes and
architectural styles.*

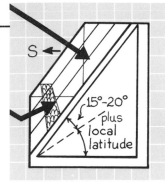

In an area with lots of sunshine, tilt can be latitude plus 15 to 20 degrees because of sunlight reflected onto the glazing from the surrounding ground.

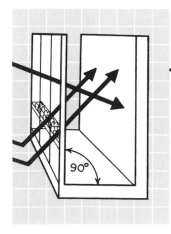

If you live at latitude 45°N or more, where there's snow on the ground for much of winter, the south wall of your sunspace can be vertical.

In general, attached sunspace retrofit design pretty much takes care of itself. For a given amount of glazing, there'll be a proportional amount of floor and wall providing thermal mass. The main decision in a retrofit is how much solar heat you want to provide for your house, and how much room is available for the sunspace. The rules of thumb for direct-gain thermal mass ratios found on pages 110 and 111 can be used.

Specific design details for all retrofit situations can't be covered in a single chapter; for more precise information, dig into the list of suggested reading at the end of this chapter.

There are several ways to provide thermal mass in a south-facing sunspace or greenhouse.

*Locate a **rock bed** in the crawl space under the floor. Warm air flows through vents in the floor to the bed, and heat is transferred from the air to the rocks, where it is stored. Use 2 cubic feet of rock for each square foot of south-facing glass.*

***Water drums** placed along the north wall should be stacked as close together as possible to allow for the conduction of heat. Use about 7½ gallons of water per square foot of south-facing glass.*

THERMAL MASS LOCATIONS

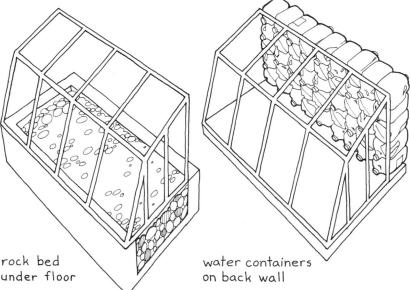

rock bed under floor

water containers on back wall

The Attached Greenhouse

The word *greenhouse* brings to mind those old all-glass hothouses with so many cracked panes, filled with plants or flowers. Conventional greenhouses were used only to grow food and flowers, and were also cold and inhospitable in bad weather.

Only when people began to attach greenhouses to residences did they become warm and cozy year-round living and growing spaces that also contribute heat to adjacent rooms. The modern solar greenhouse, far advanced over its predecessor, is designed to provide not only plenty of light for plants and/or people but some winter space heating as well.

Make sure you really need a growing greenhouse. Consider cold frames, hot boxes, or window greenhouses, for example, before a more ambitious greenhouse. Don't add a greenhouse on impulse because other people are doing it, or because it just sounds like fun. Successfully growing plants in a greenhouse requires a green thumb, and lots of work.

The amount of effort involved in maintaining a working greenhouse will depend almost entirely on whether you use it to grow food. If you plan to use your sunspace only as a pleasant sitting room and for helping to heat your house, the task will be much easier.

For a growing greenhouse, plan on spending at least an hour a day managing and working your small, glassed-in growing space. That's

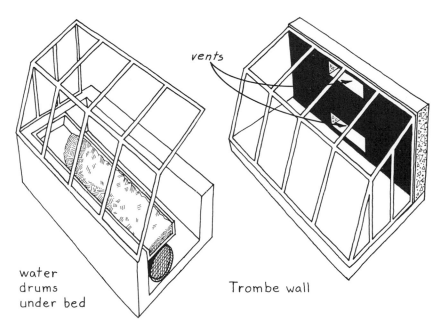

vents

water drums under bed

Trombe wall

Water drums *can also be placed under growing beds to provide heat for plants as well as for the rest of the greenhouse.*

A **Trombe wall** *has vents at top and bottom to allow air flow during the day, but the vents should be closed at night to prevent the escape of warm air from living area to sunspace.*

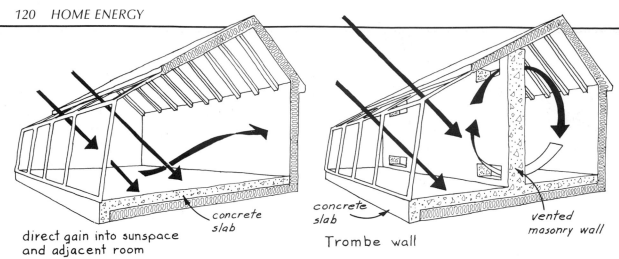

direct gain into sunspace
and adjacent room

concrete
slab

Trombe wall

concrete
slab

vented
masonry wall

*Greenhouses and sunspaces
are popular solar additions
because they make attractive,
usable living spaces. Whether
you want to use your sunspace
to grow plants, for thermal
gain, or a little of each, try to
face the sunspace as close
to due south as possible. It
should receive direct sunlight for at least four hours
each day.*

ATTACHED GREENHOUSE OPTIONS

every day, year round. If you travel a good deal, a greenhouse is
probably not for you unless you have dependable neighbors willing
to fill in while you're gone. The important point is not to take on
a greenhouse project that will be too big for you to handle. Be aware,
too, that plants require reasonable temperatures to survive and grow,
and that a single freeze in an otherwise successful year can wipe out
your crop. Design and proper insulation are much more critical in
a working greenhouse than in a sunspace.

Buying an Attached Sunspace or Greenhouse

Sunspaces and greenhouses in many shapes, sizes, styles, and
prices are available in kit form. The average floor area of kit greenhouses is about 160 square feet, but many are twice that size or
larger. Dimensions range from 8 to 12 feet wide, 12 to 20 feet long,
and 8 to 11 feet high. The larger the sunspace you can accommodate
and afford, the better. It will give you more room and provide more
heat, and probably cost less by the square foot.

Kits include framing precut to modular sizes for quick and easy
assembly. The south wall, roof, and foundation usually are insulated;
endwalls may be fully or partially insulated, all glass, or fitted with
doors and windows as desired. The greenhouse may be single- or
double-glazed, using glass, reinforced fiberglass, rigid plastic, or even
inexpensive polyethylene sheeting. Designs include shed and gambrel
roofs and straight, curved, or kneewall south glazing.

Some designs use aluminum tube framing, fitted with thermal-
break insulation to reduce heat loss. Some are designed to be detached
in summer and stored until needed, thus solving the overheating

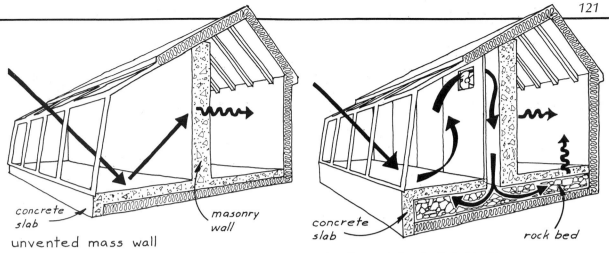

concrete slab

masonry wall

unvented mass wall

concrete slab

rock bed

air-forced heat distribution

problem. For very cold areas, insulating curtains or panels may be provided. Heat storage includes water containers or in some cases the phase-change materials described on page 106.

Greenhouse kits are generally well designed and complete with assembly instructions. However, you'll have to be familiar with local codes and other restrictions that could affect the size and style of an addition to your house. If you don't find something suitable in a kit, check for custom greenhouse builders in your area.

Building a Greenhouse

Building your own greenhouse or sunspace has several advantages, including saving money and having fun. Disadvantages include the length of time it takes to build a greenhouse and the possibility that the finished product may not be as beautiful as you had expected, or may not perform as well, either.

Unless you've done similar work, have attended a solar greenhouse construction workshop, or have access to expert advice and can handle the physical labor involved, don't take on something of this magnitude for your first project.

You should have a set of working drawings before beginning even the site preparation. Find plans that suit you, have some drawn up, or draw them yourself. Have the proper authorities check them, get a building permit, and then get busy building your greenhouse.

Backup Heating

It might be possible to design a passive house so that, even in the most severe weather, the sun would provide all the heat required, but it wouldn't be cost-effective. Collecting sufficient heat and storing it for the longest period of bad weather that might ever be encountered would require a huge amount of glazing and thermal mass. Not only would the cost be excessive, but the appearance of this all-solar house might not please everyone. So backup heat is necessary.

This is no problem for a retrofit passive solar house because the existing hot-air furnace, hydronic heating, or electric heating system can be used. So can a good wood stove or fireplace.

Your New Passive Home

A passive retrofit is a compromise design, seldom an ideal solution. But with a new house, designed from scratch with proper orientation, just the right mix of glazing and thermal mass, and insulation matched to the environment, you should have a nearly self-sufficient design. This is the time for the most sophisticated design tools and an exact blend of conservation and passive features for the most cost-effective system. Table 3-4 gives the maximum percentage of total heating needs you should aim for from solar heating for new passive houses designed specifically for the cities listed.

In addition to passive solar *heating,* a new passive house can easily incorporate the features recommended in this chapter for passive cooling. Starting from scratch, it should be no problem to include proper shading, insulation, reflective materials, landscaping, thermal mass, and ventilation. If supplemental air conditioning is needed,

Table 3-4
Annual Percentage of Passive Heating for 16 Cities

Location	Heating Degree-Days	Latitude	Solar Heating (Btu/ft^2)	% Solar Heat
Los Angeles, Calif.	1,700	34.0	53,700	100
Fort Worth, Tex.	2,467	32.8	38,200	81
Fresno, Calif.	2,622	36.8	43,200	83
Nashville, Tenn.	3,805	36.1	39,500	65
Albuquerque, N.Mex.	4,253	35.0	63,600	84
Dodge City, Kans.	5,199	37.8	58,900	72
Seattle, Wash.	5,204	47.5	42,400	52
New York, N.Y.	5,254	40.6	48,000	60
Medford, Oreg.	5,275	42.3	47,400	56
Boulder, Colo.	5,671	40.0	62,500	70
Lincoln, Nebr.	5,995	40.8	53,500	59
Madison, Wis.	7,838	43.0	44,900	42
Bismarck, N.Dak.	8,238	46.8	53,900	46
Ottawa, Canada	8,838	45.3	37,900	32

SOURCE: "Passive Solar Heating Evaluated," *Solar Age* (August 1977).

consider evaporative cooling. If you must have refrigerated air conditioning, a considerably smaller unit will do the job because of the energy-saving assist from passive cooling techniques.

Perhaps the easiest, although not the most creative, way to become a new passive homeowner is to buy one in a passive solar development. San Clemente, California, has one with hundreds of passive houses; Wayne and Susan Nichols have built several large passive solar developments in New Mexico. Having a solar house custom-built isn't as straightforward as shopping the models and making a down payment. It's more expensive, too. But many people want the satisfaction of a passive house that reflects not just energy savings, but personal tastes as well.

For those with building experience (or lots of ability and courage), a passive house built with one's own hands is the most rewarding project of all. Good passive house designs are plentiful (there are some included in a few of the books recommended at the end of this chapter), and passive design isn't that much different from quality conventional construction.

Most new passive houses incorporate the classic options—direct gain, thermal storage walls, or attached greenhouses or sunspaces.

Passive Cooling

Cooling degree-days measure temperatures above the comfort level. If your city has 3,500 cooling degree-days, it's obvious that you live in Phoenix, Arizona, or some other hot place. See Appendix A for a cooling degree-day map. However, you don't need degree-day maps to know if your house needs cooling. If so, consider these passive techniques that can help keep you cool.

Solar energy is a blessing we easily forget about when the temperature is 110°F in the shade. Solar-powered air-conditioning systems are theoretically an appropriate use of solar energy: we need heating in the winter when the sun provides the least energy, but the hotter the summer, the more sunshine is available to operate a cooling system. The possibility of cost-effective active solar cooling is described in Chapter 4, starting on page 158. Here we're going to learn how to keep cool right now with passive methods.

Keeping Heat Out

In winter you need lots of sun on those parts of your house that use it for heating: direct-gain glazing, thermal storage walls, and attached sunspaces/greenhouses. In summer keep the sun off them as much as possible. Refer to the cooling options checklist on pages 46–47

(Continued on page 126)

Custom-built solar homes, which still dominate the solar home market, borrow from both traditional and contemporary designs.

in Chapter 1 for pointers on insulation, reflective films, and other ways of keeping heat out. Common sense and conservation are as basic to passive cooling as they are to passive heating. If you've properly insulated to keep heat in, the insulation will also help keep unwanted heat out.

Monitor drapes, awnings, and outside doors carefully to be sure they're closed when they should be. Open drapes let in more sun; so do awnings that are up instead of down. Windows and doors left open in the heat of the day commit two sins: they let hot outdoor air in and cool inside air out; just the kind of cross-ventilation you *don't* want.

A white roof may absorb as little as 25 percent of the solar heat reaching it; a black roof, as much as 90 percent. So if cooling is your biggest problem, consider a light-colored roof. It's also been proved that installing aluminum foil beneath the rafters in an attic is very effective in keeping radiant heat out of the house in summer. An air space should be left at the ridge, however, so that moisture won't be trapped between roof and foil.

Landscaping is another important consideration for natural cooling. Large trees not only shade a house but also cool the air around it with the moisture they transpire. A caution, however: make sure that trees don't keep winter sun from reaching the house where it's needed. And don't let a windbreak of trees prevent cooling breezes from carrying heat away from the house in summer.

Grass and flower beds reflect less light into a house than do concrete or light-colored gravel; they also have a lot less mass to soak up heat and reradiate it to their surroundings.

In direct-gain houses, the main concern is keeping the sun from coming through all that south-facing glass into the living space. Here's where shading from a roof overhang can help. This cooling technique takes advantage of the sun's high angle during summer months. Proper roof overhang shades south-facing windows from the summer sun.

Table 3-5 gives a quick method for determining the proper overhang for both winter heat and summer shade:

$$\text{length of the overhang} = \frac{\text{window height}}{F}$$

where F is a factor from table 3-5.

For example, if your windows are 5 feet high and you live in St. Louis, Missouri, (about 39°N), the overhang should be between 5/3.4 and 5/2.5, or from 1½ to 2 feet. The 2-foot overhang will shade your windows completely until August 1; a 1½-foot overhang, only until June 1. If your present roof overhang doesn't provide window

Table 3-5
Fixed Overhang Factors

North Latitude	F*
28	5.6–11.1
32	4.0–6.3
36	3.0–4.5
40	2.5–3.4
44	2.0–2.7
48	1.7–2.2
52	1.5–1.8
56	1.3–1.5

SOURCE: Edward Mazria, *The Passive Solar Energy Book* (Emmaus, Pa.: Rodale Press, 1979).

*Select a factor according to your latitude. Higher values provide complete shading at noon on June 21; lower values, until August 1

ROOF OVERHANG GEOMETRY

Properly sized overhangs shade out hot summer sun but allow winter sun (which is lower in the sky) to penetrate windows. See table 3-5 for sizing guidelines for your latitude.

shading, awnings may do the trick. Trellises, screens, and other shading devices help, too.

If you use shade screen on your south-facing windows, put the screens outside the glass. Inside shades help, but not nearly as much as outside screens that keep the sun from getting through the glass.

Now for techniques that help an indirect-gain house. With Trombe walls or TAPs, close all vents into the house. Also be sure to vent the glazing so that heated air will escape and not heat up the absorber as much. Consider covering the glazing to keep the sun from it. Take window units down for the summer.

With sunspaces or attached greenhouses, isolate these areas by closing all doors and vents into the house. Cool your sunspace in summer by using shading to keep out sunlight. Cloth covers or plastic or bamboo blinds made for greenhouses can be rolled over the glazing. Trees and vines help, too, but they should be the kind that won't shade the greenhouse in winter. Some deciduous trees never lose all their leaves, so check before you plant.

One simple method for keeping a sunspace cool in summer is to paint the outside of the glazing white to reflect solar radiation. Use whitewash—the old-fashioned solution of lime and water—and it will gradually wash off with summer and fall rains. The last bit can be removed by hand before cold weather hits. If the sunspace needs more than vents or a small fan to keep it cool in summer, consider removing sizable sections of the glazing.

Handling Heat That Does Get In

There'll always be some solar heat that gets past your defenses, but if it's kept to a minimum, you can usually stay reasonably com-

Some low-energy options for keeping cool in summer: thermal chimneys, ceiling fans, and, for those who like to experiment, cool tubes.

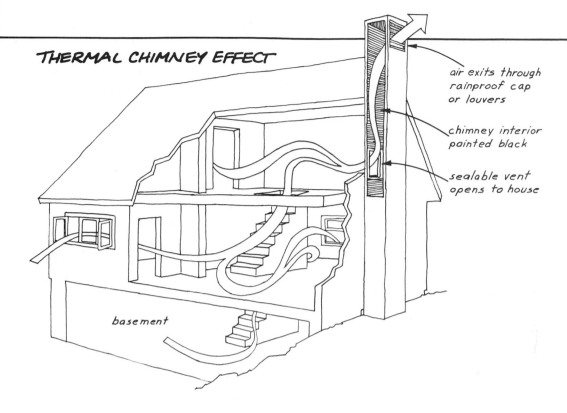

THERMAL CHIMNEY EFFECT

air exits through rainproof cap or louvers

chimney interior painted black

sealable vent opens to house

basement

---

fortable. This is especially true if your house has sufficient thermal mass for winter heat storage. Instead of *you* absorbing all the incoming heat, the walls and floors soak up a lot of it. Then when the sun is down and evening coolness comes, you can open up the windows and doors and let the thermal mass lose its heat to the outside.

If you can open windows to channel breezes through them, your house will stay much cooler—and so will you. Outside air 10°F cooler than air inside the house, and flowing at 5 miles an hour through 10 square feet of window area, matches the output of a small air-conditioning unit. If there are no breezes, make your own with an electric fan placed where air movement is needed. Whole-house fans bring in cooler outside air and get rid of hot air through the attic.

Thermal Chimney

A thermal chimney serves the same function as a whole-house fan, but without moving parts. The chimney usually sits on the roof and is the highest point on the house. It is open to the house by way of a sealable vent. The south side of the chimney is glazed to absorb heat, and the interior is usually painted black. As the sun shines on the glazed chimney wall, the air inside warms, rises, and passes out the chimney's top, creating an updraft. Cool air is drawn

A more complex alternative to a whole-house fan is the thermal chimney. It works by natural convection—air in the chimney heats and rises, pulling cool air up from the basement or crawl space. Because installation involves constructing a tower, it's most suitable as part of a new house or an addition.

up from a basement, crawl space, or lower-level shaded windows to replace the heated air leaving the chimney. The height and solar heating of the chimney speed the flow of cool air through the house.

Cool Pipes

Cool-pipe potential is based on the fact that ground a few feet below the surface remains cooler than the surface in summer. If you've visited Carlsbad Caverns, the limestone caves in Kansas City, or other caves, you know how cool it is in these places. As noted earlier, some heat pumps also tap the ground for cooling.

The method used is to bury cool pipes horizontally in the ground. One end of the piping is ducted into the house; the other is brought above the surface so it can take in outside air. As warm air rises in the house, outside air is drawn through the buried pipe and cooled. The principle behind cool pipes is an intriguing one, but the reality so far has been, for the most part, disappointing. Humidity from ground moisture can be a problem; so can reverse airflow.

Desiccant Cooling

There are probably as many people using desiccant cooling techniques as there are people who can spell desiccant. That's not very many, but the idea is presented here as another long shot that may become popular with system improvements. Desiccant is a fancy way of saying dry, and this cooling concept depends on drying the air to make it more comfortable. Early residents of Charleston, South Carolina, placed large casks of salt in warm, humid areas (salt absorbs a lot of moisture and thus dries the air). When a barrel got soggy, however, it had to be dried out before it could absorb more moisture.

Modern passive desiccant cooling developments include trays of activated charcoal, which absorb moisture from the air, plus a speedy way to dry the desiccant material. Two units are used; while one is drying the air, the other is being dried or "regenerated" by heat from a solar collector.

For More Information on Passive Heating and Cooling

Building Underground, Herb Wade, 1982: Rodale Press, 33 E. Minor
 St., Emmaus, PA 18049

 This one is for those planning an earth-sheltered house. All you
 need to know about building underground and incorporating
 passive features.

The Food and Heat-Producing Greenhouse, Bill Yanda and Rick Fisher, 1979: John Muir Publications, Box 613, Santa Fe, NM, 87501

A simple, logical guide to choosing, building, maintaining, and utilizing solar greenhouses for food production as well as supplemental heating.

The Passive Solar Design Handbook, Vol. 3, Department of Energy, 1982: Superintendent of Documents, U.S. Government Printing Office, Washington, DC 20402

This one is a must for the passive architect/engineer and will be valuable to serious amateurs. Most useful for new construction.

The Passive Solar Energy Book, Edward Mazria, 1979: Rodale Press

Generally considered the bible of passive solar design. Explains passive processes, principles, systems, and needed design tools.

Solar Air Heater, Ray Wolf, 1981: Rodale Press

Plans and step-by-step instructions for building a simple space-heating collector for a south wall.

The Solar Greenhouse Book, James C. McCullagh, 1978: Rodale Press

Excellent for the greenhouse enthusiast. This practical book covers not only attached and freestanding greenhouses but pit greenhouses and solar cold frames as well.

The Solar Home Book, Bruce Anderson, 1976: Brick House Publishing Co., 34 Essex St., Andover, MA 01810

An early one but still good. Covers basics, passive, active, and water heating.

Solar Remodeling: Passive Heating & Cooling, by the editors of Sunset Books, 1982: Sunset Books, Lane Publishing, Menlo Park, CA 94025

Probably the best bargain you'll find for passive solar basics, simple projects, and an elegant full-color showcase of completed passive retrofits.

Solarizing Your Present Home, Joe Carter, ed., 1981: Rodale Press

This giant book (688 pages) covers solar basics, water heating, space heating, and natural cooling. Loaded with drawings and projects.

Suppliers

Prices and addresses given are as of this writing; check with the supplier before sending money.

Sunspace/Greenhouse Plans and Kits

Andersen Corporation (Sales Promotion/Concept IV, Bayport, MN 55003) offers the Andersen Concept IV idea booklet, with information on how to create custom sunrooms. The booklet costs $6.95.

Brother Sun (1301 Cerrillos Rd., Santa Fe, NM 87501) offers plans for Bill Yanda's sunspace for $25. The price of plans will be credited to your purchase of Brother Sun thermal-break glazing for the sunspace.

Richard Feeney (Box 1698, Durango, CO 81301) offers plans for a sun-room using a Snowflake window wall. The plans cost $16.22 postpaid.

Jeff Milstein (Box 413, Dept. N5, Bearsville, NY 12409) offers detailed plans and instructions for a 9-foot-by-12-foot sunspace. Building experience is necessary to complete this project. The plans cost $15 postpaid.

New Mexico Solar Energy Association (Box 2004, Santa Fe, NM 87501) offers an 80-page guide for construction of straight-eave or Quonset-style attached solar greenhouses. The plans cost $6.95 postpaid.

New York State Energy Research & Development Authority (Add-On Plans, 2 Rockefeller Plaza, Albany, NY 12223) offers plans for an attached solar greenhouse and sunspace addition. A construction manual and drawings are included. The plans cost $18.

Sun-Tel (1270 Parrish St., Lake Oswego, OR 97034) offers a user's manual, construction manual, and three sheets of blueprints for a sunspace you can build with off-the-shelf materials. The plans cost $15 postpaid.

Vegetable Factory, Inc. (Box 2235, New York, NY 10164) offers plans for a 9-foot-by-12-foot sun-room. The plans cost $3.

Weather Energy Systems (Box 968, Pocasset, MA 02559) offers blueprints for a Sun Haus sunspace. They cost $55.

(The following kits are for 8-foot-by-12-foot sunspaces or as close to that size as the manufacturer makes. Prices do not include optional features.)

Abundant Energy, Inc. (116 Newport Bridge Rd., Warwick, NY 10990) offers a Radiant Room kit with double-pane, tempered glazing; laminated beams; and thermal-break framing. The wall insulation is R-11; the roof is R-22. Intermediate building skills are needed. The sunspace has a 1-year warranty with a 5-year warranty on the glazing seals. The cost per square foot is $42.

Advance Energy Technologies (Box 387, Clifton Park, NY 12065) offers the Zero Energy Room kit. The walls are made from R-40 steel-encased urethane foam panels with a fiberglass thermal break. No glazing is supplied. There is a 20-year warranty on the panels. The cost per square foot is $52.

Aluminum Greenhouses (Box 11087, Cleveland, OH 44111) offers the Everlite kit, with double-strength glazing and aluminum framing. Advanced building skills are needed. The sunspace has a one-year warranty. The cost per square foot is $19.

Brady and Sun (97 Webster St., Worcester, MA 01603) offers the Livingroom kit, with double-paned, tempered glass; arched, laminated pine beams; and exterior trim aluminum. The sunspace has a five-year warranty. The cost per square foot is $33.

English Greenhouse (11th and Linden St., Camden, NJ 08102) offers the Florex ITB kit. It includes double-paned, tempered vertical glass; a double-paned curved eave; and thermal-break aluminum framing. Intermediate building skills are needed. The sunspace has a 10-year warranty. The glazing and seals have a 5-year warranty. The cost per square foot is $71.

Green Mountain Homes (Royalton, VT 05068) offers the Solar Shed Sunspace kit, with double-paned, tempered glass in wood frames; cedar shingles; an insulated door; and vents. Intermediate building skills are needed. Defective or missing items are replaced within five days of delivery. The cost per square foot is $21.50.

Habitat (123 Elm St., South Deerfield, MA 01373) offers the Solar Room kit with double-paned, tempered glass and laminated cedar rafters with a thermal-break aluminum frame. The walls and roof have R-26 insulation. Advanced building skills are needed. The sunspace has a one-year warranty. The glazing seal has a five-year warranty. The company provides free shipping within 300 miles. The cost per square foot is $25.

McGregor Greenhouses (Box 36, Santa Cruz, CA 95063) offers the Attached Green Room 8 kit, with fiberglass glazing and redwood framing. It totals 72 square feet. Intermediate building skills are needed. The company offers a full refund if you're not satisfied. The cost per square foot is $5.

National Greenhouse (Box 100, Pana, IL 62557) offers Sunspace. The kit includes double-strength glass, aluminum framing, a ridge vent, and condensation gutter. Intermediate building skills are needed. The sunspace has a one-year warranty. The cost per square foot is $20.

Santa Barbara Greenhouses (390 Dawson Dr., Camarillo, CA 93010) offers the Lean-To kit, with fiberglass glazing, clear cedar framing, dutch doors, and vent. Intermediate building skills are needed. The glazing is guaranteed not to yellow or crack for 15 years. The cost per square foot is $6.50.

Solar Resources, Inc. (Box 1848, Taos, NM 87571) offers a kit for an air-inflated sunspace. It includes double-layered polyethylene glazing (UV inhibitor) and thermal-break aluminum framing. A blower is included. Intermediate building skills are needed. The sunspace has a one-year warranty. Your money is refunded if you are not satisfied during the first six months of ownership. The cost per square foot is $14.

Sunglo Solar Greenhouses (4421 26th Ave., Seattle, WA 98199) offers the Lean-To 1700-E. The kit has double-wall acrylic glazing, aluminum framing, aluminum bench framework, and exhaust fan. Intermediate building skills are needed. The sunspace has a five-year warranty. The cost per square foot is $14. The kit plan is also available without lumber. It costs $3 per square foot.

Turner Greenhouses (Box 1260, Hwy 117, Goldsboro, NC 27530) offers a Lean-To kit. It includes fiberglass glazing, steel framing, and storm doors. Intermediate building skills are needed. There is a 15-year warranty on the fiberglass. The cost per square foot is $6.50.

Vegetable Factory, Inc. (Box 2235, New York, NY 10164) offers a Solar Greenhouse Kit, with double-paned acrylic glazing, aluminum thermal-break framing, and storm door. Intermediate building skills are needed. The sunspace has a five-year warranty. The cost per square foot is $30.50.

Active Heating and Cooling

Passive space heating uses the natural flow of warm air and heated thermal mass like brick and concrete to distribute and store solar energy, but active space-heating systems depend on plumbing and wiring, pumps, fans, and a variety of mechanical devices and electrical and electronic controls. As a result, active heating is more complex, expensive, and maintenance intensive. For these and other reasons, active space heating is usually not as cost-effective as active water heating or passive space heating.

Despite these shortcomings, however, active space heating makes sense in some situations. For example, some houses, because of design or orientation, just aren't suitable for passive heating. Others may have some passive potential but not enough for the whole job.

An Introduction to Active Heating

Because active space heating was popular long before most people ever knew what passive solar meant, there are still more active than passive home installations. Most of these heat water for domestic use, pools, or spas, but the same kinds of collectors and associated equipment handle space heating, too, and many active space-heating systems do the job they were designed for, and do it well.

As we did in the passive chapter, we're focusing here primarily on retrofitting heating systems for existing houses. Be sure you've done all the energy conserving that you can, as well as any passive retrofitting that's practicable for your home, before you read on. If there still aren't enough Btus to keep you comfortable at a fuel cost you can afford, active heating might be what you're looking for.

What You'll Learn in This Chapter

- Whether you go with an air or a water active solar heating system will probably depend upon your present heating system because you'll want to take advantage of existing water pipes or air ducts (page 141)

- Economical active space-heating systems should be sized to provide about 50 to 80 percent of your home's total heating needs, depending upon your climate, the condition of your home, and your life-style (page 145)

- Heat storage in an active system takes the shape of a rock bin or a water tank (page 152)

- Faulty control systems have been shown to be the single greatest factor in unsatisfactory performance and poor energy economy of active space-heating systems. Because of their complexity, active systems are more trouble-prone than passive systems (page 153)

- Good, regular maintenance is the key for the proper operation and long system life of active systems (page 156)

- With just a few exceptions, passive solar makes more sense for a new home than active solar heating (page 156)

- Solar energy is most associated with heating. But when the sun is strongest, and therefore hottest, we need cooling, not heating. The key is to find a way to use strong summer sun to cool our homes. Several hundred solar air-conditioning systems are operating experimentally in the United States (page 158)

As mentioned earlier, active systems rely on mechanical components to collect solar heat and to deliver that heat to the living space and/or storage. Either a liquid or air can be drawn through the collectors to absorb heat from the sun and transfer it to storage and living spaces.

Active Heating System Operation

An active heating system has four basic components: solar collectors, heat storage, distribution system, and controls.

When an active heating system works, solar collectors, mounted where they receive lots of sun, heat air or water. Directed by a control system, blowers or pumps move the heated water or air through a distribution system of ducts or pipes to living spaces, or, if rooms don't need more heat, to heat storage. When the control system thermostat calls for heat, the distribution system carries any stored heat to the living space.

Experience has shown us that active systems are usually poor choices for new construction because of the lower cost of passive designs. But they can make sense for retrofits because design and structural house changes are usually not necessary.

This active **liquid system** for space heating uses water as its medium. Water is drawn through the collectors to absorb heat, and then pumps, activated by a control system, move it through a series of pipes to heat your house—or, if the heat isn't needed, into a storage tank. The storage tank is not a domestic hot-water heater. Because no solar space-heating system should be designed to supply 100 percent of your heat needs all the time, an auxiliary heater will be needed.

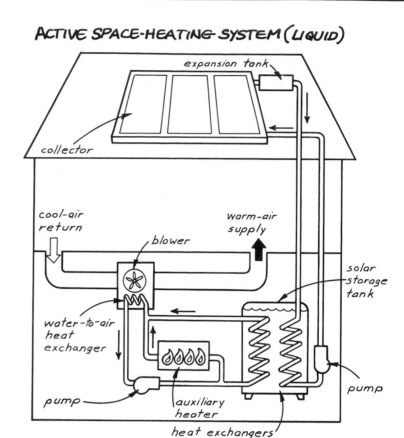

ACTIVE SPACE-HEATING SYSTEM (LIQUID)

expansion tank

collector

cool-air return

warm-air supply

blower

solar storage tank

water-to-air heat exchanger

pump

auxiliary heater

pump

heat exchangers

When integrated with backup heating, active space heating can operate in several modes: direct solar heating, heat storage, heating from storage, and backup heating. More than one operational mode can be used at a time: direct heating plus heat storage during bright, clear days, for example. Or stored heat plus backup heat in bad weather.

It's also possible to combine active and passive technologies in a hybrid system. One such approach is called active charge/passive discharge solar heating: active collectors heat water, and a distribution system pumps the water to coils in a concrete slab. Or air is heated in active collectors and blown through ducts to a rock bin beneath the floor. In either case, the warm thermal mass then slowly releases heat to the house passively.

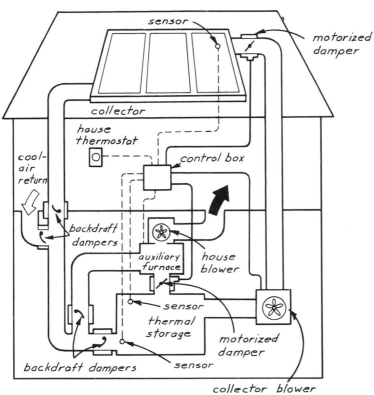

ACTIVE SPACE-HEATING-SYSTEM (AIR)

sensor

motorized damper

collector

house thermostat

control box

cool-air return

backdraft dampers

auxiliary furnace

house blower

sensor

thermal storage

motorized damper

backdraft dampers

sensor

collector blower

*The alternative to an active liquid space-heating system is an **air system.** Air flows through the collectors, is heated, and is then moved by blowers through a duct system into your living space. If the heat isn't needed, dampers reposition and send the warm air into storage (usually a pebble bed, but phase-change material may also be used). When heat is being supplied from storage, cool room air enters the bottom of the storage unit and, once warmed, exits from the top. If the heat from storage is insufficient, the auxiliary furnace automatically turns on.*

Choosing Your Options

The basic components mentioned above are required in all active heating systems, but some options are available for more flexible system design. As you've already learned, solar collectors can be either air or water types. Similarly, heat storage can be masonry, water, or phase-change materials. And the heat distribution system components will depend on the collectors used and the storage medium they heat.

Your choice of collector and storage options will depend mostly on the heating equipment in your house now. If you have radiators or baseboard hydronic heating, the logical options are liquid col-

SOLAR AIR-HEATING MODES

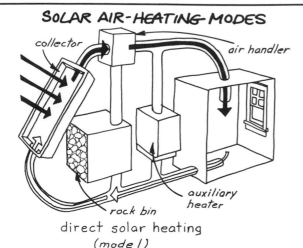

direct solar heating
(mode 1)

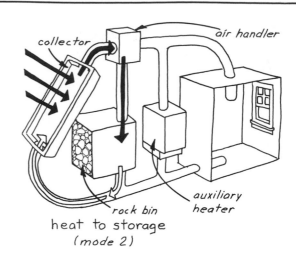

heat to storage
(mode 2)

The operating modes of a solar air-heating system with a rock thermal storage bin and a backup heating system are controlled automatically by thermostats and air handlers (blowers and dampers). If the house thermostat indicates that the house needs heat on a sunny day, solar-heated air from the collector bypasses the rock storage and moves directly to the house (mode 1). When the house no longer needs heat, the hot collector air is diverted to storage (mode 2). If the house requires heat on a very cloudy day or at night when there is no hot air available from the collector, hot air from the rock bin is moved to the house (mode 3). When the house needs heat and neither the rock storage nor collector can supply it, the backup (auxiliary) heater takes over (mode 4).

lectors and a water-storage tank. This will facilitate system integration, minimize heat losses, result in a more efficient system, and simplify backup heating. If you already have a radiant heat floor slab with integral piping fed with conventionally heated water, you should also use water collectors.

If you have a gas furnace and hot-air heating, you might want to consider air collectors. As in the above examples, the reason is for compatibility with existing heating equipment. Of course, it's also possible to pump water through fan coils for hot-air heating. In the early years of active heating design, heat exchanger applications were often carried to extremes. But why use heat exchangers, with their added expense and loss of efficiency, when you don't have to? If you're now using electric resistance heating, you can choose almost any active heating options, since electric heat can be used as backup for either air or water systems.

As with passive designs, it's cheaper to use some heat from conventional fuels than it is to build a huge active solar system sized to take care of several days of heat storage. So don't get rid of your gas or oil furnace or heat pump unless it's worn out and can't be used for backup heating. Also use as much of the existing air ducts or water pipes as you can. Careful planning will let you tie the new active system into your old heating plant and delivery system.

Fitting It All In

You'll need a sizable area of south-facing roof or wall for mounting collectors. You'll also need considerable space for heat storage; either a water tank or a rock bin. Preferably, heat storage should be inside the house. This is particularly true for a rock bin so that duct

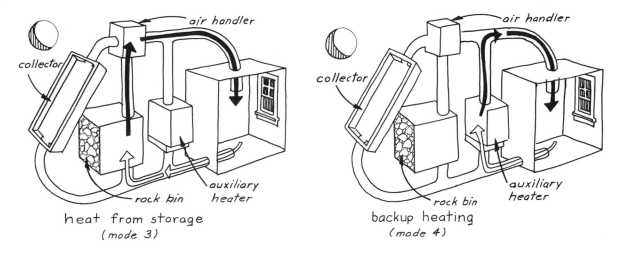

heat from storage
(mode 3)

backup heating
(mode 4)

runs can be kept short. If absolutely necessary, the water tank can be outside but should be buried for insulation and appearance.

The Economics of Active Heating

It's wise to do a realistic economic check before committing yourself to an active heating project. For starters, its cost will far exceed that of a domestic hot-water system; make sure it will pay back the cost. Active heating costs are higher than those for passive heating, because of first cost and also because of maintenance. A later section in this chapter covers maintenance; don't omit this item from your financial balance sheet.

The dollars-and-cents of solar water heating was covered extensively in Chapter 2; read it again and evaluate active heating for your particular situation. Active space heating starts at a disadvantage because the collectors are generally used only during the winter months, rather than all year as solar water heating is.

Table 4-1
Component Options for Active Heating

Collector	Heat Storage	Heat Delivery
Water	Water tank	Radiators; baseboard heating; fan coils (to heat air)
	Concrete slab	Metal or PVC pipe
Air	Rocks or gravel; containers of water; phase-change materials	Natural convection; forced air

THE RIPP HOUSE HYBRID SOLAR SYSTEM

The Ripp house in Troy, New York, is an example of a hybrid active/passive home. Heated air, collected from a second-story skylight, is drawn into a vent at the top of the stratification tower by a variable-speed fan. The fan automatically turns on when the temperature upstairs gets above 70°F, and moves the warm air under the concrete slab and into the tunnels through the cinder blocks. The blocks absorb some of the heat and slowly release it to the slab. When the still-warm airstream reaches the end of the blocks, it is directed through another vent and into the living area.

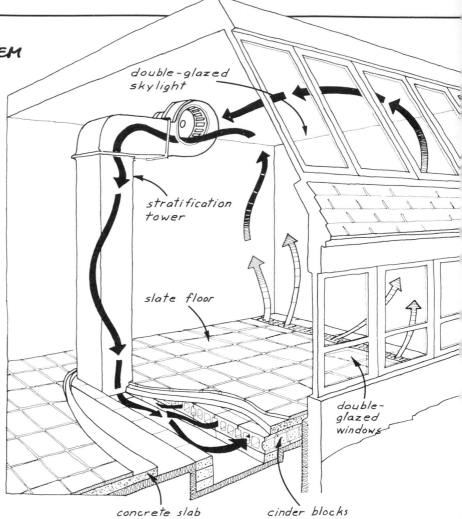

double-glazed skylight

stratification tower

slate floor

double-glazed windows

concrete slab

cinder blocks

If you can use the collectors for pool heating in spring and fall, their cost-effectiveness will be greatly increased. And if you can make active solar cooling work for you despite its high cost now, you'll be using the collectors year round. As the section on active cooling points out, however, that technology can't yet compete with conventional cooling. So don't jump the gun on passive cooling.

Designing Your Active Heating System

Right up front: unless you have considerable experience in heating design, it's a good idea to have a professional help plan your active system. Make sure he or she is truly professional by checking references and, if possible, talking with past clients.

With or without professional help, the first step is to determine how much collector area is required. This will depend on how much heat you need and the availability of sunlight in your area. If you're only interested in heating a room or one wing of the house, a few collectors may do the job. To heat a large portion of your house, it'll be necessary to add an expensive array of collectors.

As with a passive heating project, if you're retrofitting active solar to your present home, you'll have to adapt active heating components to the house and its existing heating system. However, an active retrofit is usually easier than a passive one because there's no need for adding a sunspace or greenhouse, or installing additional south-facing windows. Active solar collectors can be mounted on the roof fairly easily, and except for heat storage in the basement or elsewhere, the interior of the house can remain pretty much as it is. There'll be no glare from large expanses of glass, and no exterior glazing or cutting of vents for Trombe walls or thermosiphoning air panels (TAPs).

Design Tools

There are many methods of calculating collector area for a given floor space in a given climate and house with certain construction and insulation characteristics. Turn back to page 108 in Chapter 3 for some ideas because sizing tables and RULES OF THUMB aren't that different for passive and active systems. Also check the listing of books at the end of this chapter for a sampling of design methods.

If you're interested in computer design techniques, see Appendix C for a variety of programs for home computers or hand-held calculators. If you hire a professional designer, he'll most likely use one of these, or a program of his own. Aiming for 100 percent of your house's heating from an active system isn't practical. Rather, follow this RULE OF THUMB:

A practical active space-heating system should be sized
to provide 50 to 80 percent of your home's total
heating needs.

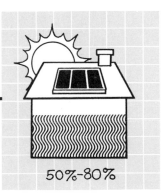

50%-80%

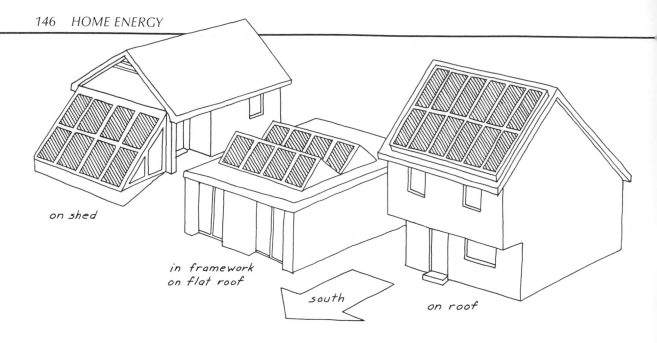

on shed

in framework
on flat roof

south

on roof

ACTIVE SOLAR COLLECTOR LOCATIONS

Rather than try to make a solar designer of you in this book, table 4-2 gives active collector area worked out by Westinghouse a few years ago.

Table 4-2 gives a very rough idea of collector area required in different locations to provide 50 to 80 percent solar heat for a 1,600-square-foot house. You can tell from the table that if you live in Wisconsin (severe climate), you're going to have a bigger active heating project than someone in Santa Maria (mild climate). This table gives only a very rough idea because there are two key factors that significantly influence these numbers: your house's heat loss and the insolation available in your area.

Tables like these generally are designed to work for houses of typical construction and insulation. These have heat losses of about 10 Btu per square foot per degree F. If your house is better than this (a tight, well-insulated house loses only about half that amount of heat), you can use less collector area.

Degree-days and insolation aren't the whole story, either. How hard does the wind blow? Can you tolerate 68°F, or must you keep the thermostat higher? Do you turn the heat down—or off—at night?

Using the Solar Index

There's a very simple way to find out the heat requirements of your house, and that's by calculating your costs for heating it. Table

Although the roof is usually the most convenient and out-of-the-way location for mounting collectors, you have other choices that may better suit you. A handful of options are shown here: on a shed; mounted on a flat roof; mounted on a south-facing roof; mounted vertically on a wall; mounted on a pitched roof (stand-off unit).

in vertical
mounting
on wall

in stand-off
unit on pitched roof

2-4 in Chapter 2 used the Solar Index and its computer-derived data to size collectors for water heating in various cities. This information can be adapted for sizing space-heating collectors, too. Here's how.

Carefully check your utility bill for the coldest month, and compare it with a summer month when you used no space heating at all.

Table 4-2
Collector Areas for a 1,600 Ft² Residence

Location	50% Solar Heat (ft²)	80% Solar Heat (ft²)
Mobile, Ala.	120	240
Santa Maria, Calif.	160	220
Atlanta, Ga.	240	820
Wilmington, Del.	440	1,120
Madison, Wis.	1,200	not enough roof!

SOURCE: Samuel Glasstone, *Energy Deskbook* (Washington, D.C.: Technical Information Center, U.S. Department of Energy, 1982).

A Quick and Dirty Estimate of the Solar Heating Value of Your Active System

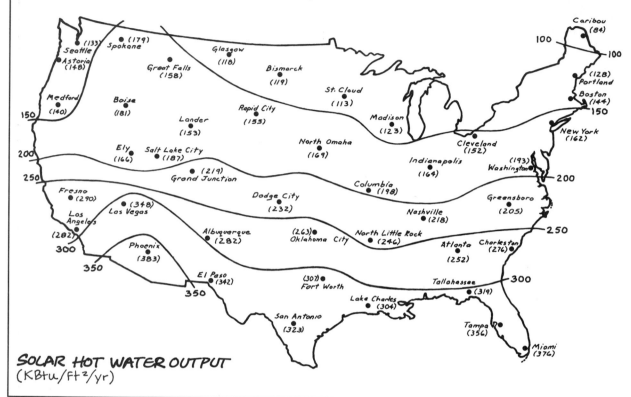

SOLAR HOT WATER OUTPUT
(KBtu/Ft²/yr)

If all other things are pretty much equal (you didn't have company during one bill and did during the other, or some other such variation), the difference in cost should fairly well indicate the added energy used for heating.

If you use gas for hot-water and space heating, the arithmetic will be simple. If you have an oil furnace for space heating but use an electric water heater, you'll have to find out the differential costs per Btu between them (oil is generally much cheaper than electricity).

Having done all this homework, you can calculate the ratio of space-heating cost to hot-water cost. Let's say it comes out three times as much energy used for space heating as for hot water. So if you need 60 square feet of collector for hot water, you need about 180 square feet for space heating. If the cost ratio was 6 to 1, you would need 360 square feet.

Like the map used for solar water heating on page 60 in Chapter 2, this contour map lets you make a quick estimate of the output of an active space-heating system. It works exactly like the solar water-heating map: the numbers next to contour lines and city names represent the solar output (heat actually delivered to the living space) in terms of thousands of Btus per square foot of collector area per heating season. You can pick a number that is closest to your area and multiply it times the collector area you're thinking of installing. You'll end up with a huge number, in the millions of Btus, so just divide the result by 1 million to get the result expressed in MBtus. That number can be multiplied by your present heating cost (in $/MBtu), which you figured out in the box "How Much Does Your Home Energy Really Cost?" on page 22. What you'll get for all your effort is the equivalent dollar value of the solar heat you'll get from your system. In other words:

$$\frac{\text{MBtu/ft}^2/\text{yr} \times \text{ft}^2 \text{ of collector area}}{1{,}000{,}000} \times \begin{array}{l}\text{present heating cost in} \\ \text{\$/MBtu} = \text{dollar value of solar} \\ \text{heating}\end{array}$$

If you live near the 100-KBtu contour line, the system you install will deliver about 100,000 Btu per square foot of collector area per heating season. Thus, 200 square feet of collector will provide 20,000,000 Btu, or 20 MBtu, of space heating (200 × 100,000 = 20 MBtu). If you were heating with oil at a cost of $11 per MBtu, you could expect a return of about $220 a year. Of course, before you actually make the investment and begin any major work like this, you should consult with a solar engineer or architect who can give you a more thorough and detailed analysis of solar performance for your area.

Having established the required collector area, check to see if you have enough properly oriented roof space. (Orientation is covered in detail in Chapter 2.) In all the cities in table 4-2, except Madison, a conventional, south-facing roof should provide sufficient collector space for active heating. In Madison, you'd have to add wall-mounted collectors and maybe ground-mounted ones, too, for even 50 percent solar heat.

Making Active Heating Work

Once you've got a rough estimate of the system size you think you need, you can use the quick and dirty method in the box here to determine the dollar value of the solar heat you'll get from such a system. These calculations can also be found in Chapter 3, for passive system estimates.

Components

In order to get a sense of your active heating options and see how they can link up with the heating equipment already in your house, let's look at the components in more detail. We'll start with liquid-type collectors.

Collectors

Instead of one or a few flat plate collectors as in domestic water heating, a space-heating system requires a sizable array. Murphy's Law works with a vengeance when applied to collector connections, as suggested by this RULE OF THUMB:

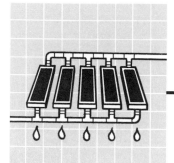

Five interconnected collectors are about 25 times as likely to leak or otherwise malfunction in a given situation as one collector is.

So, within reason, use the largest collectors you can buy or build. Connecting a large number of small collectors results in far more plumbing fittings plus a greater possibility of leaks, trapped air, and other problems. A 4-by-12-foot collector provides 48 square feet of area with only two plumbing fittings. Four 2-by-6-foot collectors also give 48 square feet total area, but require eight interconnections plus the headaches of providing series and parallel flow paths, air venting, and other plumbing problems. Large collectors may be cheaper by the square foot, too; another reason for going big.

At first thought it seems logical to install air collectors rather than water collectors for a space-heating system, since it's air that you want to heat in the house. Furthermore, air collectors won't freeze in cold weather, or leak a liquid in any kind of weather. First impressions can be misleading, however.

For example, air collectors can leak air, and air leaks are a lot harder to detect than water leaks. If you do opt for air collectors, you'll have to be very sure they're airtight. Be sure that all ducts are airtight, too.

Air collectors have another problem. A much greater volume of air than water must be pushed through the distribution system to deliver a given amount of heat from collector to living space; the noise can be a nuisance.

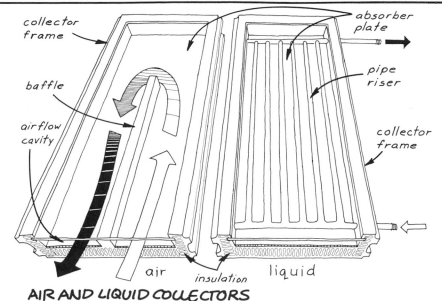

AIR AND LIQUID COLLECTORS

When it comes to choosing between air and liquid solar collectors, there are advantages and disadvantages to be weighed. Many people choose **air collectors** because they are simple to build on site. They are less likely to have freezing or corrosion problems, and they can be incorporated easily into existing forced-air heating systems. The drawbacks are air leaks, both internal and external. An internal leak releases hot or cold air from one part of the system to another at the wrong time or in the wrong way. An external leak releases solar-heated air to the outside.

Liquid collectors have been popular because liquids hold more heat in a smaller amount of space than does air, and thus, smaller pipe diameters are used. Liquid systems can be easily incorporated into hot-water space-heating systems. Water leaks are the big concerns in a liquid system, along with corrosion and freezing.

Regardless of which type you choose, your best bet is to use commercially manufactured collectors. New-generation collectors are well built, long-lasting, and very efficient. If you are in a very cold climate, invest the extra money in double-glazed collectors to prevent heat loss. Any collector will work, but good grade commercial collectors will work better—and last longer.

Table 4-3
Comparison of Liquid and Air Collectors

Liquid	Air
Advantages	
More selection of commercial collectors	Simpler design
Generally more efficient	No corrosion problems
Can be combined with domestic hot-water or pool-heating systems	Compatible with forced-air heating
Pipes small & easily insulated	Cannot freeze
	Lower cost
Disadvantages	
Freeze protection needed	Leaks hard to find
Corrosion problems	More energy required for delivery system
Leaks harmful	Noisier delivery
Overheating problems	Ducts & heat storage bulky
Installation & heat-storage costs generally higher	

Heat Storage

In passive heat storage, the floor and/or walls can serve as heat-storing thermal mass since the sun shines directly on them. In an active system, heat from the collector must be delivered through ducts or pipes to a rock bin or water tank. Because of their size, such storage units are generally placed in a basement, although sometimes they can be installed in large closets if the floor will support their weight. For example, an early active air system in Denver stored gravel in large-diameter cardboard tubes placed in the living room and painted to harmonize.

The volume of heat storage required depends upon the area of the collector. With a small collector, there's not much reason for installing a large heat-storage bin or tank because most or all of the collected energy will be used just for daytime heat. A properly sized collector allows the storage of part of the heat for use during night-time or bad weather. Too much thermal mass will never get warm enough to be of use. Heat-storage volume will also depend on the material you use.

Two RULES OF THUMB provide guidelines for sizing storage:

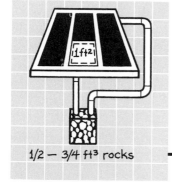

1/2 — 3/4 ft³ rocks

Figure on about ½ to ¾ cubic foot of rock per square foot of collector.

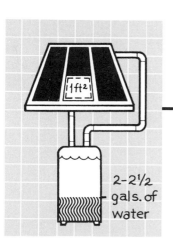

2-2½ gals. of water

Water has considerably more heat-storage capacity than do rocks, so, cubic foot per cubic foot, you will need less than half the size rock bin for the same size collector. That means about 2 to 2½ gallons of water storage per square foot of collector.

Heat Distribution

In addition to collectors and heat storage, you'll need a distribution system of ducts and/or pipes, plus fans to move air or pumps to circulate liquid. Minimize the length of ducts and pipes by installing collectors and heat storage as close together as practicable. This simplifies ductwork, plumbing, and insulation; it also requires smaller fans or pumps.

Installers of conventional heating systems generally don't realize how much heat can be lost between collectors and the space to be heated. For an active heating system to work efficiently, ducts and pipes must be well insulated. And if you live in an area where freezing is possible (and that's almost everywhere), you'll need freeze protection. This is covered in detail on page 76 in Chapter 2.

Controls

A control system is intended to:

Help the system collect as much solar energy as possible

Keep the house comfortable at least cost

Minimize the use of backup heating

Minimize fan and pump operation

Operate freeze-protection equipment

Prevent system malfunction or breakdown

Extend the life of components

A variety of controls are needed to properly operate an active heating system, including the following: differential thermostats, temperature sensors, damper controls, pump and fan controls, and a master control panel.

Performance evaluations of hundreds of active heating systems show that faulty control systems are the single greatest factor in unsatisfactory performance and poor energy economy. Even in commercially installed systems, controls are sometimes inadequate and/or improperly adjusted. Take plenty of time on the design of your control system, the purchase of components, and proper installation and check-out procedures.

Backup Heating System Integration

The impracticability of making a solar collector/heat storage system large enough to take care of the worst possible weather conditions has been explained. As with practically all renewable-energy applications, backup heating of some kind must be integrated with the active heating system. This is generally a minor problem, however. Just keep using your existing gas furnace, electric heat, oil stove, wood stove, or whatever. You won't have to use it nearly as much, so it'll last a lot longer.

Two New England active solar retrofits for space heating. The one below is a hot-air system; the house on the right uses circulating hot water.

Problem Areas

Because problems occur in direct proportion to system complexity, active systems are much more trouble-prone than passive systems are. And while control systems are the big offenders against proper performance, they're not the only ones. Here are frequently cited faults:

Substandard installation
Collector shading
Improperly sized collectors/
   heat storage
Uninsulated/leaky heat storage
Active system poorly integrated
   with existing heating system

Poor heat delivery: improper
   size fans, pumps, ducts;
   insufficient dampers; poor
   airflow through rock bin
Corrosion of metal parts
Homeowner ignorance of proper
   system operation

The last problem can be eliminated by acquiring full knowledge of how your system is supposed to operate. Proper maintenance will ensure continuing satisfactory operation.

Who'll Do the Work?

Unless you're an expert yourself, hiring a reliable solar heating contractor is the quickest and generally the surest way to a successful active retrofit system. The "shopping around" advice given in earlier chapters applies equally well to active heating projects. Be sure you have a good design, quality equipment, and all the expert advice you can get. If tax credits are available, keep accurate cost records and apply for them. You can afford a better system that way.

While there are some really fine commercial collectors now on the market, building an active heating system can be a rewarding do-it-yourself project, *if* you know what you're doing. And if you've built one collector, you can build several (and probably do a better job on each one). Doing it yourself, you can also size the collectors to your needs and to the configuration of your roof.

You may save money doing it yourself, but don't underestimate the time an active retrofit project will take. Start in the spring if possible so you can complete the work before the snow flies. You'll also need good working drawings, and have them checked for necessary permits. You don't want a ton of collectors falling through your roof, water flooding your house, or an electrical fire sending the whole thing up in smoke.

Maintenance

Because an active heating system uses solar collectors that are subjected to extreme environmental conditions, and because such a system has many moving parts and electrical and electronic equipment, it will require good maintenance for proper operation and long service. Use common sense, follow instructions, and you'll have fewer problems.

- Keep fans and pumps in working order by regular lubrication and prompt replacement of worn parts
- Immediately repair leaks that develop—whether liquid or air
- Antifreeze must be changed periodically as recommended by the manufacturer for proper operation and protection of plumbing
- Check the control system periodically and maintain it for proper operation. This includes the various sensors and mechanisms that turn fans and pumps on and off and operate freeze protection
- Keep collector glazing reasonably free of dust and dirt; replace or repair broken glass as soon as possible

As with any kind of maintenance, the best approach is checkups and preventive maintenance on a regular basis to prevent serious problems from ever occurring. Most manufacturers warrant their products and stand behind them. You should be given a checklist with purchased equipment and told how and when to do what. During summer, for example, unused liquid-type collectors must be kept from reaching high temperatures and boiling. This is generally done by draining the collectors and properly venting them. It may also be wise to shade or cover them to keep them cooler.

In winter a preseason checkup—preferably one performed by a qualified serviceman—is necessary. The water or antifreeze pipes must be refilled without air bubbles. Control operation must be checked and necessary adjustments made. Freeze protection must be tested for proper operation; duct dampers and heat registers should be adjusted for even heating throughout the house.

Active Heating in New Homes

Designing a *new* house for active solar when passive is possible rarely makes good sense. More than likely, anyone who chooses an *active* solar new home over a passive one can't calculate properly,

Roseville, California, boasts one of the few solar subdivisions in the country. Each house is equipped with active roof collectors for heating and cooling.

or needs a new designer. On the other hand, contemporary subdivision homes in Roseville, California, use active collectors on roofs to provide radiant space heating because cost estimates showed that this approach was cheaper than the passive designs originally planned. And the active system provides these houses with a bonus—summer cooling as well, with the same equipment.

The Lyndonville, Vermont, house that *Rodale's New Shelter* March 1983 cover hailed as a new generation in solar technology is a new home, designed from the ground up for a roofful of active water collectors heating a concrete floor slab, which then passively heats the house.

The house works remarkably well in an 8,000-degree-day environment, using only a wood stove for backup heating. The active charge/passive discharge system is being used by other builders across the country. Radiant heat from the floor slab eliminates the need for a complex system of ducts, fans, and dampers. There are only four moving parts in the heating system: a pump, an air-purging valve, a one-way check valve, and a ball valve.

This new-generation active design embodies five general principles. All are covered throughout this chapter but summarized for emphasis:

1. Get your priorities straight
     Use all the conservation techniques first; then use
     active solar

2. Keep it simple
     Stick with the necessary basic components and don't gussy
     up the system with dozens of moving parts

3. Accept no substitutes
     Use only simple, reliable, high-quality components

4. Install with care
     Make your installation as high-quality as the components
     you select

5. Take care of your active system
     This includes maintenance, as well as proper use

Active Cooling

Space cooling requires only a fraction as much energy as space heating, perhaps 3 to 4 percent of total energy use in the United States, compared to the 20 to 25 percent used for heating. However,

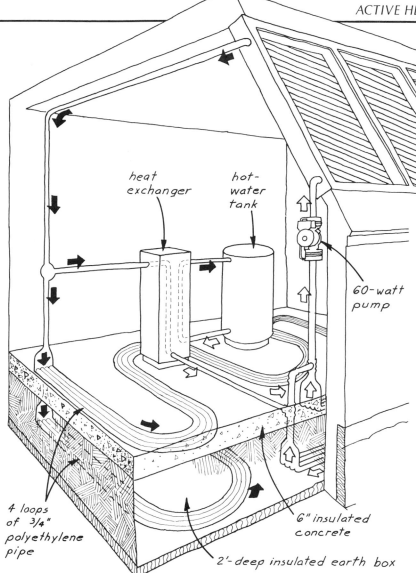

heat
exchanger

hot-
water
tank

60-watt
pump

4 loops
of 3/4"
polyethylene
pipe

6" insulated
concrete

2'-deep insulated earth box

THE STARR HOUSE HYBRID SOLAR SYSTEM

The Starr house, in Lyndonville, Vermont, is a hybrid solar home. It combines the simplicity of passive design with the architectural flexibility and comfort control of active designs.

When the sun is shining, an electric controller switches on the pump, which pushes a water/antifreeze solution into the collectors for heating. The warmed antifreeze solution is then fed into three pipes. One runs to the heat exchanger, which preheats the tap water supply. The second pipe runs to a "header," which feeds the liquid to the four pipes embedded in the concrete slab. A second header sends the remaining antifreeze through the four loops in the earth box.

As the antifreeze circulates through the loops, it gives up its heat to the concrete and earth (which provide thermal storage and slow release to living areas), and then returns to the pump, where it is joined by the antifreeze that ran through the heat exchanger. The pump then sends the antifreeze back to the collectors for reheating.

for those living in the furnacelike environment of the Southwest, or the hot and steamy South, cool summer comfort is more important than winter warmth—and costs far more. Population is another driving force behind the development of active solar cooling technology.

The warmer Sunbelt and West are growing much more rapidly than the North; cooling is thus increasing its share of energy use.

Solar energy is most associated with heating, for we all know the sun's rays are hot. This is fine in cold weather, but when the sun is strongest and therefore hottest, we need cooling, not heating. The key, of course, is to find a way to use the strong summer sun to cool our homes rather than heat them. Research and development on solar air conditioning has been going on for more than a decade; several hundred systems are operating in the United States, and another thousand or so around the world.

Solar Cooling on the Way?

Years ago there were lots of refrigerators that operated on a gas or kerosene flame, and gas-fired air-conditioning systems are still in use. These absorption-type units were modified some years ago for solar cooling applications: hot water from solar collectors was substituted for the gas flame. Latest development in the solar absorption cooling field is the invasion of the United States by Japan's Yazaki Company, which has built a large assembly plant in Texas and aims to capture the market.

Pioneer solar scientist John I. Yellott has been operating a Yazaki 2-ton absorption air conditioner in his Phoenix, Arizona, home for six years. A total of 720 square feet of Yazaki flat plate collectors provide the 200°F water that takes the place of the old gas flame in absorption refrigeration systems.

Yellott's solar air conditioner supplies more than 90 percent of the cooling needed in the fiery desert environment (more than 3,000 cooling degree-days), all the winter heating needed, and ample hot water the year round. The system saves $1,200 a year but, at a cost of about $15,000, will require a very long payback period.

Several American companies are also in the race for the active cooling market. Among them is American Solar King, the leading manufacturer of flat plate collectors in the United States. Its entry is an active solar desiccant cooling system (see Chapter 3 for passive desiccant cooling), to be marketed in 1985 for about $6,000.

For More Information on Active Heating and Cooling

The Complete Handbook of Solar Air Heating Systems, Steve Kornher, 1984: Rodale Press, 33 E. Minor St., Emmaus, PA 18049

The 720 square feet of solar collectors on John I. Yellott's house in Phoenix provide summer cooling, winter heating, and year-round hot water.

A complete step-by-step guide to building several types of air-heating collector systems, including window and wall-mount units, full wall and roof systems with heat storage, systems for domestic water heating.

Homeowner's Guide to Solar Heating, by the editors of Sunset Books, 1978: Sunset Books, Lane Publishing, Menlo Park, CA 94025

This Sunset book is still useful for basics, design, and installation ideas for retrofit and new homes. Color gallery of homes.

The Solar Decision Book, Richard H. Montgomery and Walter F. Miles, 1982: John Wiley & Sons, 605 Third Ave., New York, NY 10158

An excellent book that covers new and retrofit solar houses. Features detailed active, passive, and hybrid approaches in 20 well-organized decision chapters.

The Solar Home Book, Bruce Anderson, 1976: Brick House Publishing Co., 34 Essex St., Andover, MA 01810

This pioneer book on solar homes is still a very useful guide to active solar heating and cooling.

Solarizing Your Present Home, Joe Carter, ed., 1981: Rodale Press

Recommended in Chapter 2 and Chapter 3 for solar water heating and passive retrofit work respectively, this book also covers active heating and cooling for existing houses. Good for do-it-yourself projects.

Suppliers

Prices and addresses given are as of this writing; check with the supplier before sending money.

Thermosiphoning Air Panel Plans and Kits

ALR Energy Products (RD 1, Box 415, Belvidere, NJ 07823) offers the solar hot-air panel kit for a 35-inch-by-77-inch collector. It includes frame, absorber, fiberglass glazing, fan, and thermostat. The collector is a window unit. The kit costs $299 plus shipping.

Havlick Snowshoe Co. (Box 508R, Gloversville, NY 12078) offers plans for a hot-air solar heat collector. This unique TAP carries the heated air through a 4-foot stovepipe. The unit is wall-mounted, but warm- and cold-air ducts enter through the existing window. The collector is 42 inches by 86 inches. Plans and instructions cost $5 postpaid.

Mother Earth News (Box 70, Hendersonville, NC 28739) offers plans for a window-mounted solar collector. It fits any standard window, and can be vertical- or angle-mounted. The plans cost $10. Materials cost about $80.

New Mexico Solar Energy Association (Box 2004, Santa Fe, NM 87501) offers nine pages of detailed plans and instructions for a window-mounted TAP. The unit fits into a window like an air conditioner. The plans cost $1.75 postpaid.

Office of Human Concern (Box 756, Rogers, AR 72756) offers a solar air heater (TAP) kit. The unit is 4 feet by 8 feet and is designed for wall mounting. Component parts may be purchased as a kit or individually. The manual costs $5.95 postpaid, and the kit costs $130 plus shipping.

Our Solar System, Inc. (6704 Stuart Ave., Richmond, VA 23226) offers a solar air panel (TAP) kit. The unit is 3 feet by 7 feet and designed to heat about 250 square feet. It has an aluminum frame and absorber, glass glazing, insulation, fan, and thermostat. The TAP can be built at the company's workshop or ordered assembled. The kit costs $325 at the workshop; $400 mail-ordered plus shipping.

Solar Usage Now, Inc. (Box 306, Bascom, OH 44809) offers the DIY Air Panel kit. The panel is 35 inches by 77 inches. The kit includes frame, plastic glazing, fan, and thermostat, but not the absorber. This is a window unit. The kit costs $99 plus shipping.

Sunkeepers Thoughtful Solar Applications (1488 Sandbridge Rd., Virginia Beach, VA 23456) offers complete plans and instructions for building a solar window box collector. It fits into a sash-type window like an air conditioner, and is designed to heat a 12-foot-by-12-foot room during the day. The plans cost $6.

Urban Solar Energy Association (595 Massachusetts Ave., Cambridge, MA 02139) offers a Solar Wall Collector Packet, a collection of introductory articles on the basic principles and applications of TAPs and other wall-mounted collectors. The packet costs $2 postpaid.

Heating with Wood

Wood heating must have a lot going for it for 15 million households to be using it. With a good, big wood stove, you can heat your whole house. You can use smaller wood stoves to heat portions of your house, or to serve as backup for passive or active solar heating systems.

There are two obvious advantages of heating with wood. From a financial standpoint, it can save a considerable amount of money. And there are obvious aesthetic advantages to be enjoyed from using wood for fuel. There's also a disadvantage you should consider.

In spite of its popularity, wood heating does involve much more work than conventional heating does: cutting and hauling wood, the mess and dirt of handling it, shoveling out and disposing of ashes. If you haven't used wood before, be sure to consider these very real chores and inconveniences along with the advantages as you look at wood heating.

Types of Wood Stoves

You have a choice of dozens of wood stoves, but the field can be narrowed right away by dividing wood stoves into two basic types: airtight stoves and those that aren't airtight. Airtight stoves can deliver efficiencies of about 60 percent. Stoves that leak air are in the 30 to 40 percent efficiency range. So buy an airtight stove if you're serious about heating your house and a cheaper type only if you just want to build an occasional fire.

Franklin

Modern Franklin stoves differ quite a bit from Ben's original design. Basically they're open stoves, practically freestanding fire-

What You'll Learn in This Chapter

places, certainly not airtight. Most Franklins are equipped with a log or coal basket so air can come up from beneath for better combustion. The doors can be kept open for viewing the fire, or closed for more efficient burning.

Potbelly

The potbelly is probably the most recognizable of all stoves. Once it warmed passengers in train depots, hotels, and general stores. With lots of radiating surface to crowd around, the potbelly could—and still can—warm lots of people at once. These stoves have two air inlets, the top one providing "secondary combustion" for better efficiency. A good potbelly stove can be reasonably airtight.

Circulating

A circulating wood stove offers performance close to automatic heating. Some are equipped with thermostats, and some have blowers

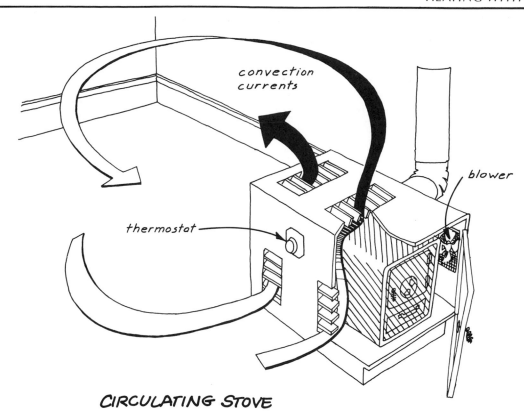

CIRCULATING STOVE

to improve air circulation. The most distinctive difference is the double-wall construction of the stove, allowing heated air to circulate between the walls. Because the outer wall of a circulating stove gets nowhere near as hot as that of a regular wood stove, it's the safest type to be around and a good choice if you have small children. The circulator does have a couple of drawbacks: the surface is too cool to cook on, and its no-nonsense practicality just doesn't give it the charm of some other wood stoves.

Scandinavian

The Scandinavian designs are so named because the original airtights were built in Scandinavia; today many of them come from places like Korea, Taiwan, and the United States. The distinguishing features of good Scandinavian types are airtightness and scientific baffling inside the stove for more complete combustion. The better Scandinavian stoves are made of cast iron, with doors ground to fit

A circulating wood stove comes close to operating automatically. Cool air enters through bottom vents, is heated, then, with the help of a blower, circulates between the stove walls and out through the top vents. The vent damper is controlled by a thermostat that keeps the amount of heat entering the room constant. Convection currents are set up in the room, and as the heated air cools, it drops and reenters the stove.

Because of the double-wall construction, the outside of the stove stays cool—a consideration if you have children in the house.

A Russian stove is a style of masonry stove. The principle of a masonry stove is to put as much chimney space as possible into the living area. This accomplishes two things: it extends the flue length, allowing the fuel to burn fast and clean; and the large amount of internal masonry acts as thermal mass.

The flue in this Russian stove/fireplace has a series of baffles, creating a greatly increased flue path that extracts the maximum amount of heat from the fire. The extended chimney stores and then gradually releases heat into the room. The owners of this stove burn wood only 4 hours a day, yet the bricks radiate heat for 24 hours.

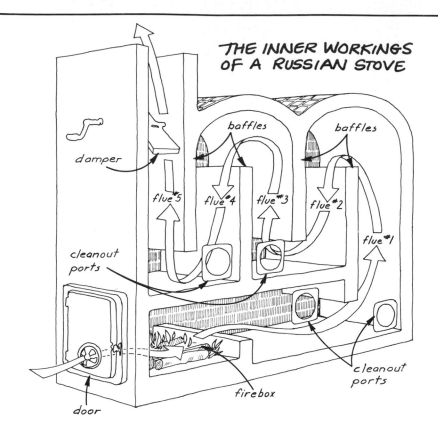

THE INNER WORKINGS OF A RUSSIAN STOVE

tightly and firebrick linings for long life. Disadvantages include high cost (such as overseas shipping for the imported ones) and the chance that nuts and bolts and stovepipe openings aren't the same standards and sizes used in the United States.

Russian

We'll wind up the description of general stove types with the Russian stove, also called the masonry wood-burning stove. You won't find these on display at your local stove shop; if you want one, you'll either have to find a specialist to build it for you, or, if you're up to the job, locate a set of plans and make one yourself. The Russian stove's distinctive feature is its long, mazelike airflow path. It also has a secondary air inlet, or blast gate, near the ceiling level that is intended to keep the flue clean.

The meandering flue channels hot air and burning gases over lots of thermal mass. Don't expect quick heating, but the big brick stove

The Russian stove, shown here and in the drawing opposite, represents one style of masonry stove. Because of their thermal mass, Russian stoves provide plenty of heat long after the fire has died.

will radiate heat long after the fire has gone out. The hot fire burns cleanly, releasing few pollutants and forming little creosote. Clean-out plugs can be built into the stove to make the flue accessible for cleaning.

Catalytic

The catalytic combustor is the newest development in the wood stove and fireplace insert business. First marketed in 1979, the device is truly revolutionary. Results from a number of testing laboratories show that pollutants—particulate emissions—are reduced about two-thirds. And because of more complete burning of the wood, heating

A catalytic stove is so named because of the catalytic converter built in to reduce pollution and boost operating efficiency. The converter in a well-designed stove should be wrapped in a stainless-steel sleeve to prevent cracking. It should be placed low enough in the firebox to receive hot gases from the burning wood. Other important elements to look for in a catalytic stove include a pre-heated secondary air source 4 to 6 inches below the combustor to mix oxygen with the gases; a baffle plate that forces the gases to pass through the combustor, positioned to separate the firebox from the heat-exchange chamber; a safety bypass to allow smoke to escape the stove if the combustor becomes clogged; and, over the combustor, a thick metal shield to prevent the top of the stove from overheating, and to reflect heat back into the catalyst.

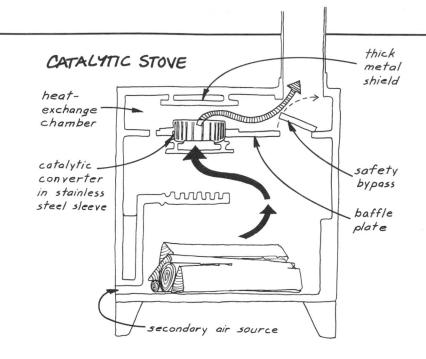

CATALYTIC STOVE

heat-exchange chamber

catalytic converter in stainless steel sleeve

thick metal shield

safety bypass

baffle plate

secondary air source

efficiency increases 15 to 25 percent. When added to a stove already in the 60 percent range, the combustor results in a very high efficiency.

The catalytic combustor is similar to your automobile's catalytic converter or smog device designed to reduce engine pollutants. A typical catalytic combustor for a wood stove is a flat, honeycombed metal disk about 6 inches in diameter, coated with a rare-metal catalyst such as palladium or platinum. The converter chemically lowers the temperature at which wood smoke burns, causing the smoke to combust. This releases extra heat into the house while consuming pollutants that would otherwise go up the chimney.

Kerosene Heaters

Kerosene heaters, mostly Japanese imports, are all over the place these days, and for good reason. Kerosene costs a little more than fuel oil, but this extra cost is balanced out by the fact that these heaters are reasonably priced, portable, convenient, clean, and don't require any installation. If pleasure is part of your argument, however, a wood stove (at least one with a glass door) may win out. You can't get much enjoyment watching kerosene burn. There are also some problems with using kerosene heaters: for example, the danger of spilling fuel, or tipping a heater over and starting a fire. And since the heaters are unvented, they do put exhaust fumes in the air.

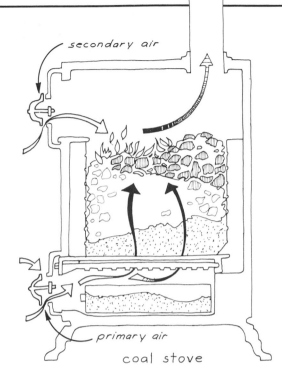

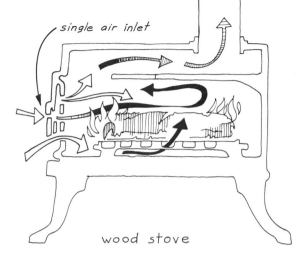

wood stove

coal stove

WOOD STOVE AND COAL STOVE AIRFLOW

Coal Stoves

Coal contains more Btus per pound than wood or charcoal; that's why so much is used in industry. The important thing about coal is that it burns a lot hotter; you'd better not try to use it in a stove designed to burn only wood, unless you retrofit the stove. Check with the manufacturer to see if the company sells retrofit packages for its stoves. If the manufacturer can't supply the parts, it's best not to attempt a retrofit.

A combination wood/coal stove, either ready-built or retrofitted, is not as efficient as a single-fuel stove. Your best bet may be to buy a coal stove and burn wood in it when you want, instead of trying to convert a wood stove to a coal burner.

You'll have to provide more clearance around a coal burner than a wood stove because it gets so hot. Otherwise, the guidelines given for installing and maintaining wood stoves apply to coal stoves as well. Instead of creosote, as in wood stoves, soot can build up inside

(Continued on page 174)

The airflow needed to burn coal safely and efficiently in a coal stove is very different from that required in a wood stove, and it is the main reason why coal can't be burned in a wood stove without important stove modifications.

A **coal fire** *depends on two sources of airflow: a primary updraft from below the grate through the coal bed, and a secondary flow that enters in front of the coal bed, near the top of the fire-box. Both air flows exit out the back or top of the stove. The primary draft controls the rate of burning, while the secondary airflow aids the stove's combustion efficiency.*

A **wood fire** *can be fed oxygen from a single air source that enters in front of the stove. The airflow can be downward, crossward, or upward, and it exits through the back or at the top of the stove.*

How Much Will Wood Heat Cost You?

The energy value and cost (in $/MBtu) for wood is difficult to pin down because the variables involved can yield widely differing results. Different kinds of wood can have energy contents (MBtu/cord) that vary by as much as a factor of two, and the moisture content of one type can change its energy content by as much as a factor of seven! When you add a wide range of market prices for cordwood, plus the range of efficiencies that are possible in different wood-burning appliances, you could go crazy trying to find out if wood heat is really worth the investment.

Tables 5-1 and 5-2 don't attempt to give you that kind of bottom-line information, but they do give you some ballpark figures. Table 5-1 lists different types of wood and their MBtu-per-cord energy content with different moisture levels in the wood. You can see right away that the drier the wood, the more potential energy it contains.

Go to table 5-1 and pick the type of wood you are likely to be buying and the moisture content at which it will be burned. (A 20 percent moisture content is typical for thoroughly "seasoned," or air-dried, split wood.) This gives you the MBtu-per-cord factor. This is multiplied by the efficiency of the wood burner you are using, which is found in table 5-2; take a number in the middle of the efficiency range if you don't know just what your efficiency is. Multiply MBtu per cord times efficiency and divide the result into the unit cost ($/cord). This gives you a dollars-per-MBtu number that you can compare with the cost per MBtu of other energy you may now be using for heating; the box "How Much Does Your Home Energy Really Cost?" in Chapter 1 gives you some ballpark comparisons. To repeat, that's

$$\frac{\text{unit cost}}{\text{MBtu/cord} \times \text{efficiency}} = \text{dollars/MBtu}$$

For example, say that you could buy green white oak for $75 per cord. If you waited and air-dried it for about 12 months or more, it would increase its energy content from 4.6 MBtu per cord to 22.7 MBtu per cord by reducing its moisture content to 20 percent from 70 percent for green or freshly cut wood. Basically, the longer you let your wood dry, the lower the moisture content will be, down to a lower limit of around 20 percent. Achieving a 5 percent moisture content could be done only with an intensive drying system such as a solar-heated wood-drying "kiln."

Assume that the dried oak is used in a wood stove that operates at an average 45 percent efficiency. The cost per MBtu for wood heat in this case is:

$$\$75 \div (22.7 \times 0.45) = \$7.34/\text{MBtu}$$
$$(\text{unit cost}) \div (\text{MBtu/cord} \times \text{efficiency}) = (\text{dollars per MBtu})$$

That's relatively low-cost heating fuel. But what if you had dallied and didn't get your wood in time to thoroughly dry it? If you burned it green, you might be green with anger at yourself for using the fuel improperly and expensively:

$$\$75 \div (4.6 \times 0.45) = \$36.23/\text{MBtu}$$
$$\text{(unit cost)} \div (\text{MBtu/cord} \times \text{efficiency}) = \text{(dollars per MBtu)}$$

If the stuff burned at all, you'd be better off by far using electric heat full blast, because even this most expensive of energies doesn't cost that much.

Table 5-1
Heating Value per Cord of Different Wood at Different Moisture Contents (MBtu)

Wood	5% Moisture	Air Dry 20% Moisture	35% Moisture	50% Moisture	70% Moisture (green)
Ash	24.8	20.0	14.8	10.8	4.1
Aspen	15.5	12.5	9.3	6.8	2.5
Beech	27.0	21.8	16.1	11.8	4.4
Birch	26.4	21.3	15.8	11.5	4.3
Douglas fir	22.3	18.0	13.3	9.7	3.7
Elm	21.3	17.2	12.7	9.3	3.5
Hickory	30.5	24.6	18.2	13.3	5.0
Maple	23.1	18.6	13.8	10.0	3.8
Oak, red	26.4	21.3	15.8	11.5	4.3
Oak, white	28.1	22.7	16.8	12.3	4.6
Pine (eastern white)	16.5	13.3	9.8	7.2	2.7
Pine (southern yellow)	25.4	20.5	15.2	11.1	4.2

Table 5-2
Typical Efficiencies of Wood-Burning Appliances

Appliance	Efficiency Range (%)
Masonry fireplace	−10–10
Manufactured fireplace with heat circulation & outside combustion air	−10–10 15–35
Nonairtight stove	15–40
Fireplace stove	20–40
Freestanding fireplace	20–40
Fireplace inserts	35–55
Supplement furnaces	40–60
Central furnace	40–75
Circulating stove	45–55
Radiant stove	45–70

Coal stoves, such as the one shown here, differ in design from wood stoves. A coal stove has two air inlets instead of one, has a smaller grate, and usually has an ash drawer to allow the dumping of ashes while the stove is burning.

a coal burner and chimney, so they should be cleaned regularly. A coal burner should have an ash drawer to allow the dumping of ashes while the stove is burning.

Pollution Control

Pollution from wood burning has to be considered, too. A typical colonial New England home burned about 30 cords of wood to get through a hard winter. With better stove design and good house insulation, a few cords generally suffice today, but we also have far more homes burning wood now.

Oregon, a state filled with forests and hundreds of thousands of stoves that burn wood from them, has a steadily growing pollution problem. A study by the Oregon Graduate Center found that a wood stove operating on a typical winter day produces 10 times the carbon monoxide that an automobile does on a 50-mile trip. Furthermore, chemical analysis has found a hundred different chemicals and compounds in wood smoke—14 of them known cancer-causing agents.

As early as 1978, residential wood burning was found to be responsible for between one-third and one-half the respirable particulates in the air. These are pollutants that can lodge in the lungs and increase the risk of diseases such as asthma, bronchitis, and emphysema. In early 1983, Oregon's House of Representatives passed a law requiring pollution standards for new stoves and compliance with those standards by 1986 in 13 counties.

The Bonneville Power Administration, under the gun from the federal government to do weatherization programs in the Northwest, had to stop providing this service to homeowners with wood stoves. Pollution inside wood-heated homes was so high that making the homes tighter with more weather stripping, caulking, and storm windows could raise indoor pollution to dangerous levels.

The Pacific Northwest isn't the only area troubled by wood smoke. Wyoming is considering similar measures, since in some cities 70 percent of residents burn wood. Missoula, Montana, has prohibited "visible smoke" from wood stoves when air inversions occur. Vail, Colorado, restricts the number of wood stoves per building, and other limitations are placed on wood burning. Beaver Creek, another Colorado ski resort, has a computerized monitoring system that signals stove users when they must stop burning wood!

Across the country in Boston, a study by the Electric Power Research Institute turned up cancer-causing agents produced by wood fires. Where wood was burned, indoor pollution levels were up to 20 times higher than outdoors.

Solving the Problem

Catalytic stoves are one answer to wood stove pollution, but they have some problems of their own. Sales were brisk in the first year or so after introduction of catalytic combustors but have tapered off since then. Price is undoubtedly a factor, since a catalytic stove costs considerably more than a conventional one. A second factor is that the catalytic combustor itself may last only a year or two. Its lifetime is further shortened by the burning of aluminum foil, paper printed with colored inks, and similar catalyst contaminants. When the combustor wears out, a new one must be installed or the old one sent back for replating with the catalytic material.

Add-on or retrofit catalytic combustor packages for existing wood stoves have been marketed at prices as low as $120. Another development is the addition of catalytic combustors to fireplace inserts, an improvement that should correct problems of creosote buildup in both firebox and chimney. Improvements continue to be made, and if the lifetime of present combustors can be extended to 10 or more years, the future of the devices seems to be assured, particularly as new anti-pollution laws continue to be passed.

There are cheaper ways to reduce pollutants and improve the efficiency of your stove. Barometric dampers are an example. They're installed on the stovepipe back of a stove to provide the right amount of combustion air. A weight keeps the damper closed when the stove isn't burning, but when air is flowing through the stovepipe, the damper opens in proportion to the amount of draft. Properly adjusted, the barometric damper automatically provides the right amount of air for good combustion, and hot enough exhaust gases to keep pollutant emissions low.

Wire mesh filters can also reduce pollution from a wood stove by straining out unburned particles in the smoke. The filters are installed in the lower end of the stovepipe, close to the stove outlet. They can be rotated to lie at a right angle to or parallel to the smoke flow, the same as a stovepipe damper. Because the filters trap particles, they must be cleaned regularly to maintain their effectiveness.

Wood stove pollution can also be reduced by using common sense and good judgment. Results won't be as dramatic as with a catalytic combustor, but a cleaner and more efficient heating system will result from following some simple rules:

Use well-seasoned firewood; hardwood if possible (it burns cleaner than softwoods or wet or green woods)

Buy the smallest stove that will heat the space you want heated. Don't buy a large stove and burn a slow fire most of the time; this causes more pollutants

When starting a fire, let it burn with lots of draft for about 15 minutes before closing the air inlets and damper. And don't starve the fire for air afterward, either. Go outside once in a while and see if your chimney is smoking excessively. Open air inlets on the stove if it is

Don't close the stove up tight at night to save money. Give it some air to cut down on pollution in your chimney flue and in the air outside as well

How Wood Burns

Wood and other carbohydrates are very stable fuels: they won't burn well unless heated to a temperature at which rapid oxidation takes place. In the case of wood, this combustion temperature is about 600°F. Unless you can afford to burn kiln-dried wood, your firewood will contain moisture that must be evaporated as the wood is heated. When dry kindling heats the wood to 212°F, the boiling point of water, the moisture-evaporating first stage of the burning process is complete.

When the temperature of the wood reaches about 600°F, it ignites. This pyrolysis stage breaks down the wood into charcoal and flammable gases. The wood isn't yet burning efficiently, however, since much of the gas goes up the flue without burning completely. At this low temperature, the gases also mix with water vapor to form creosote, which coats the inside of flue pipes. For heating efficiency and safety, the gases should be burned as completely as possible.

Coaling, the complete burning of the charcoal, begins at about 1,000°F. When the temperature climbs another 100 degrees to about 1,100°F, and sufficient oxygen is present, the gases burn completely. This produces a very efficient fire in the stove, with gases so hot they burn up resinous material that would otherwise become creosote.

How Stoves Work

When the three stages described above are complete, combustion is going on in your stove. For safe, efficient heating, it's important to control this combustion, or rate of burning. Controlled combustion in turn heats your home to the temperature you want and helps to save fuel. Adjustments of the air inlet and flue damper control the burning rate.

Draft controls and air inlets, as you might guess, control the amount of air entering a stove, and therefore determine whether a fire burns slowly or quickly. They are much more effective on an airtight stove than on one that leaks air. With a proper system of baffles inside the stove to direct the airflow, a good wood stove does a better job of burning those gases generated in the pyrolysis stage of combustion. Proper travel of air through the wood stove also makes for more efficient transfer of heat from stove to living space.

Heat Recovery

A perfect wood stove would extract all the heat from its fuel and not let any of it go up the chimney. But there's no perfect wood stove any more than there is a perpetual-motion machine. Such a stove would contradict a basic physical principle: some heat must go up the flue to create the draft that keeps the stove burning.

While a certain amount of upward heat flow or draft is necessary, however, most stoves send far too much heat up the flue. This

INTERNAL AIRFLOW PATTERNS

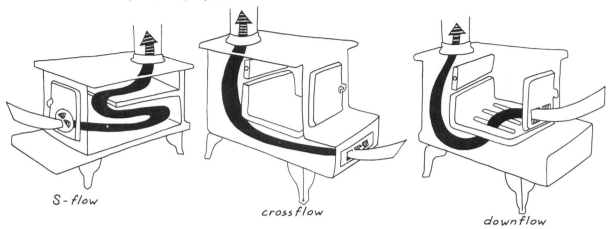

S-flow   crossflow   downflow

*When you burn wood in a stove, you want to create an airflow that will allow for maximum combustion and heat transfer. The number and size of inlets and the number of baffles affect airflow patterns. The best pattern is the **S-flow**, which occurs in Scandinavian or foreign airtight stoves. In these stoves, the number of baffles forces the air to loop under and around within the stove. The result is that the gases stay in the stove for a long period of time before they exit, transferring a large amount of heat to the stove itself, and allowing for relatively complete combustion.*

is where good design comes in: for the stove itself, and for heat-recovery devices that attach to the hot stovepipe.

Chapter 1 describes the new 95 percent efficient gas furnaces. The temperature of the exhaust gas from them is only about 100°F, much cooler than the gas sent up the stack of a conventional furnace. This is because almost all the heat has been squeezed from the burning gas inside the furnace. While it is not possible to have wood stove flue gases this cool because creosote would form, a good deal more heat can be extracted from wood stove exhaust gases in a similar manner by flowing the air over as much burning wood and hot metal as possible before letting it escape through the flue.

Look at the drawing of various draft designs. In diagonal drafts, updrafts, crossflows, and downflows, the air follows a pretty straight path from inlet to outlet. In an S-draft, the air travels farther and thus transfers more heat in the process.

An extra air inlet in some stoves also recovers heat from exhaust gases. The inlet works like the afterburner on a jet engine; fresh air mixes with the hot gases leaving the burning wood, and the secondary combustion gives more heat. These are old ideas, but they've been fine-tuned for better performance as more is learned about stove design.

As most heating systems do, wood stoves heat by radiation, convection, and conduction. A hot stovepipe also radiates and conducts heat to help warm the house. There are several ways you can make the stovepipe give off more of its heat.

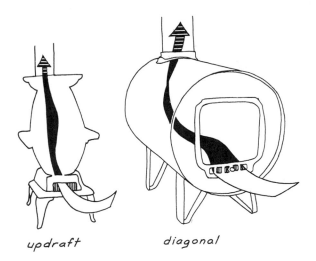

updraft   diagonal

Simple fins, cut from sheet metal and rolled around the pipe, provide more hot metal surface to radiate and conduct heat into the room. More efficient stovepipe heat exchangers are available commercially, including fan-equipped ones designed to fit into the stovepipe. Room air is blown through them to pick up heat (but not flue gas) from the stovepipe. U.S. Department of Energy tests show that such heat exchangers can improve wood stove efficiency up to 20 percent.

These and other heat-recovery devices are most effective on stoves that aren't very efficient to begin with. That's because stovepipes on nonairtight stoves get hotter than those on airtight stoves. In fact, heat recovery may be unnecessary or even cause problems on very efficient airtight stoves that send a relatively small amount of heat up the flue. As pointed out earlier, cooling the flue too much can cause excessive buildup of creosote and increase the risk of fire in the flue.

What to Look For in a Wood Stove

The more you think about wood stoves before you actually start shopping, the better the chance of bringing home a design that meets your needs.

The other stoves shown here offer less-efficient airflow patterns. **Crossflow** *and* **downflow** *stoves rely on secondary air inlets for complete combustion. In a crossflow stove, the primary and secondary air inlets are aligned, one in front of the stove, and one behind the coals. In a downflow stove, the secondary inlet is below the coals. In both stoves, the secondary air inlets are supposed to ignite unburned volatiles that would otherwise go up the flue. But many times the volatiles cool before the secondary air gets to them, and the inlets then act like leaks.*

The potbelly stove has an **updraft** *that enters at the base of the stove, passes through the wood, and exits out the top. Gases spend only a short time in the stove, and there is little heat transferred. The same is true for the barrel stove, which has a diagonal airflow. Air enters at a low inlet and is carried up diagonally through the wood and out the upper back.*

If your stove is not airtight and sends too much heat up the flue, increasing the radiant surface area of the flue or extracting heat from the gases in the flue will increase the stove's efficiency.

*A **stovepipe oven** is double-walled and cylindrical. Flue gases circulate between the walls, heating the oven to as high as 400°F. You can use a stovepipe oven (so long as your stove is not an airtight) for baking, the purpose for which it's designed, or as a heat exchanger simply by leaving the oven door open.*

*   **Add-on fins** are metal rings that slip around the flue pipe, increasing the radiant area.*

*A **fan-powered heat exchanger** is a metal box with a series of tubes in it. It extracts heat by allowing flue gases to flow through the box and around the tubes, making the box hot. Then a fan blows the hot air into the room. These types of heat exchangers should not be used with airtight stoves because they will cool the flue too much, which may result in excessive creosote buildup.*

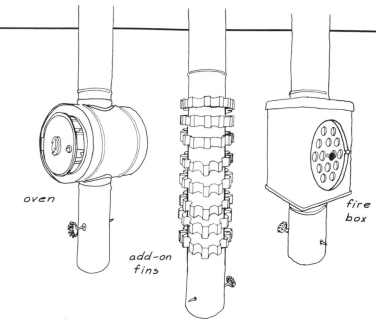

oven

add-on fins

fire box

STOVEPIPE HEAT EXCHANGERS

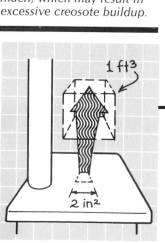

1 ft³

2 in²

Heating Capacity

Some people don't worry much about sizing. They just buy the biggest stove they can afford, on the theory that anything never used to its limit will last longer. This approach may give you a huge stove that impresses friends, but you may also have an inefficient heating system. A better approach is to size your wood stove as you would an oil or natural gas furnace—by matching its heat output to the volume of space you want to heat.

John Vivian, author of *The New, Improved Wood Heat* (Rodale Press, 1978), gives some stove-sizing RULES OF THUMB:

For each 2 square inches of radiating surface on a run-of-the-mill steel stove, 1 cubic foot of room air can be heated. A better-built steel stove might heat 1½ cubic feet for each 2 square inches of surface.

An efficient 100-pound cast-iron stove will heat two average-size rooms; an ordinary 100-pounder will heat one room.

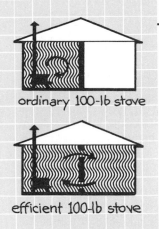

ordinary 100-lb stove

efficient 100-lb stove

A 500-pound stove will store enough Btus to give off heat for about four hours after all the wood is burned.

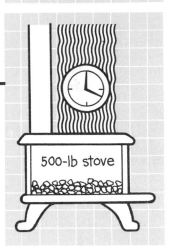

500-lb stove

Efficiency

Heating efficiency is that percentage of the energy content of wood that's converted to heat in the house. Manufacturers often quote combustion efficiencies for their stoves; the percentage of heat they extract from the wood. Using this standard, the efficiency of wood stoves on the market is claimed to range as high as 70 percent. Actual tests of the ratio of heat delivered to a room as a proportion of heat in the wood work out to about 60 percent for the best stoves.

Check two important characteristics when you're checking the efficiency of a stove. One is airtightness and the other is heat-transfer capability. To be really airtight, a stove has to be welded (if it's a steel stove) or cemented (if it's cast iron). You may be able to check for airtightness with a flashlight (hold the flashlight outside the stove, and look inside). If light leaks through, so will air. Check the tightness of the door by putting a sheet of paper between it and the stove body and closing the door. If the paper slips out easily when you pull on it, you don't have a tight door.

You can't check the heat-transfer capability of a stove on the sales floor, but you can estimate how well it should perform. The more radiating surface there is in proportion to the volume of the stove, the more heat will be transferred to the living area.

One of the most efficient wood stoves you can buy is an airtight stove such as the ones originally designed in Scandinavia; its S-flow air pattern extends the hot air's path in the stove proper. A variation on this design is an airtight stove with a heat exchanger. The smoke from the fire rises and circulates around the baffles and heat exchanger before exiting. Here, as in the traditional S-flow design, heated air remains in the stove long enough to transfer large amounts of heat to the baffles and stove walls, and to ensure relatively complete combustion.

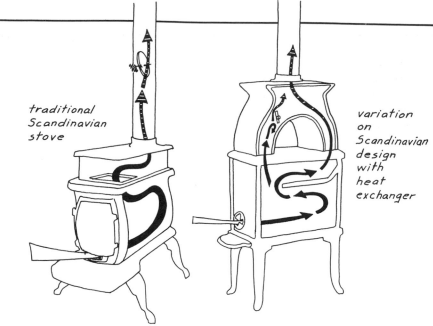

traditional Scandinavian stove

variation on Scandinavian design with heat exchanger

SCANDINAVIAN STOVE AIRFLOW

Steel plate and cast iron both radiate heat very well; that's why most stoves are made from them. The thickness of a stove's steel skin determines how quickly and how long it will transfer heat. Thin steel heats a room more quickly than does thick steel but also cools off more quickly; thick steel takes longer to begin transferring heat but continues to do so long after the fire is out.

Combustion Control

An airtight wood stove with adequate, tightly fitting air inlets and a stovepipe damper provides excellent control of combustion, or the rate of burning. You can open the inlets all the way and make the fire roar, or shut them down tight and have the fire burn all night. This second option, by the way, is an inefficient use of wood, and can cause creosote problems as well. Here's why.

If you shut the air inlets all the way and go to bed, by morning you may have a stove filled with a good approximation of charcoal. Charcoal is made by pyrolyzing, or burning wood without sufficient oxygen. Charcoal keeps most of the heat energy of the wood, so you won't get much heat from your fire during the night. What you will get is lots of creosote in the stove, stovepipe, and flue. So give the fire some air during the night, and it will burn cleanly and efficiently.

This is one point in favor of the nonairtight stove: you just about can't starve it for combustion air.

Other factors also affect how much control you have of your stove. A heavy steel or cast-iron stove gives more-even heat than a thin steel stove. And because a larger stove holds more wood, it heats evenly over longer periods of time without restoking.

A good circulating wood stove generally offers the highest degree of controllability. Equipped with automatic thermostats and dampers, such stoves come closest to the precise temperature control of natural gas furnaces. A circulating wood stove with automatic fan control also distributes heat more evenly throughout the space to be heated.

At the other end of the scale, with the least heat control, is the freestanding stove or fireplace with a fire much like that in an Indian tepee. It just burns, and there's not much you can do about control except close the damper in the stovepipe.

Cast Iron, Steel, Soapstone, and Tile

Cast iron has achieved its reputation for good reasons: it resists warping under repeated heating and cooling and is less likely to burn out than thin sheet steel. Cast-iron stoves promise a long enough life to become valuable antiques. Stovetop cooking on a cast-iron stove is like using a cast-iron griddle, and that's another advantage.

With its many worthwhile advantages, cast iron also has some problems that you should consider. Cast iron stores lots of Btus but it also takes lots of time to start radiating that heat. So if you want quick heat, a thinner steel stove may be for you. Consider, too, that for all its solid look, cast iron is brittle and can crack when subjected to sudden changes of temperature. That's why a new cast-iron stove should be broken in, or tempered, by several small fires before larger fires can be burned safely.

Finally, airtightness is achieved in a cast-iron stove by using furnace cement between the separate parts. Heat makes this cement brittle, and after a time it cracks and permits air leaks. Once in a while you'll have to recement the joints, and this isn't a simple half-hour project.

Steel stoves come in more varieties than cast-iron stoves do. There are cheap stoves made from thin sheets of steel, and there are durable steel stoves made from plates as thick as cast iron. Each type has advantages and disadvantages. Sheet metal stoves are light and portable; they also heat up quickly, and that can be an advantage.

On the other hand, some sheet metal stoves are thin enough to warp after repeated heating and cooling cycles. Warping looks bad and can break welds and cause air leaks. Light sheets of steel also rust or burn out after a while. Heavy steel stoves sacrifice low cost and fast

Wood Heating Checklist: Answer the Following BEFORE You Buy a Wood Heater

Are you physically able to operate a wood heater (have the strength and wherewithall to haul in firewood)?

Do you have a place for a wood heater (wood stove, fireplace, insert or the like)?

Is there room for proper wall clearances and a fireproof base?

Will you install an airtight, high-efficiency heater?

Do you have a good chimney, or a place to put one?

Will you have the chimney approved by an expert mason or by your fire department?

Is firewood that is high in fuel value and reasonably priced readily available?

If you cut your own wood, do you have the skills and equipment to do the job safely?

Do you have a convenient and dry place to store wood?

Can you easily dispose of ashes?

How will you heat your house when you're away for some time?

Do local laws or building codes restrict the use of wood heating?

Will a wood heater affect your insurance rates?

Will you install a chimney fire alarm and smoke alarms?

Will you regularly inspect the chimney for creosote buildup?

Will you thoroughly clean the chimney, stove pipe, and heater at the end of the heating season?

Does every member of the family know how to properly and safely operate the heater?

Does each person know how to prevent fire hazards and what to do in a fire emergency?

---

heating but offer the lifetime, durability, and heat-transfer characteristics of cast iron.

Soapstone is a popular alternative to cast iron or steel for stoves. Soapstone is a soft stone composed of talc, chlorite, and sometimes magnetite, and the name comes from its soapy feel. Soapstone stores lots of heat and won't rust or crack.

Tile is the traditional material of many European stoves. They are beautiful stoves with some of the qualities of soapstone stoves. Be prepared to pay a premium price for either soapstone or tile, however.

Other Features

In addition to basic performance, you should be on the lookout for features that effect maintenance, ease of operation, and usefulness. These include such things as replaceable metal liners, firebrick inserts

The wood-heating boom brought with it a proliferation of new stoves that proved to be, for the most part, much better designed and more efficient than earlier ones.

for your stove's firebox, and easily accessible wood-loading doors and ash-removal bins.

Although most new wood stoves are painted with high-temperature flat enamel finishes (black in most cases), there are still a few stoves that must be repainted periodically with stove black to keep them looking good. Consider carefully whether or not you want that painting chore. Circulating stoves usually are finished in porcelain enamel (or tile), and, because the outer jacket isn't exposed to the high temperatures of the fire, they seldom need repainting.

Installing Your Wood Stove

Heat rises, so don't expect a wood stove upstairs to warm the basement. Neither will a heater in a downstairs room heat one directly above it unless there's a passageway so the warmed air can travel easily. Installing a stove near a stairwell is a good idea. Ceiling vents in the room where the stove is can also heat the space above it.

You may find that one stove can't do an adequate job of heating your whole house, especially if the house is long and narrow. You then have the options of heating only part of the house or using more stoves. In a pinch, circulating fans may move enough warm air to help one stove heat more space.

Existing Fireplace and Chimney

With an existing fireplace and chimney, installing a wood stove can be simply a matter of finding a neat, safe way to get the stovepipe into the opening in the chimney. Before using an existing chimney, however, check it (or have it checked) for the following:

General condition (no loose or missing bricks, mortar still solid)

Tile or fireclay flue ⅝ inch thick or more

Opening large enough for stovepipe

Proper metal or fireclay thimble into flue

If your home has no chimney, you'll have to build one. Because a masonry chimney is an expensive undertaking, a retrofit wood stove project generally calls for a prefabricated metal chimney. Such a chimney can be double or triple wall, with insulating material between its walls. The insulation is important; without it, the air circulating between the chimney walls can cool flue gases from a wood fire so much that

LOCATIONS FOR A WOOD STOVE

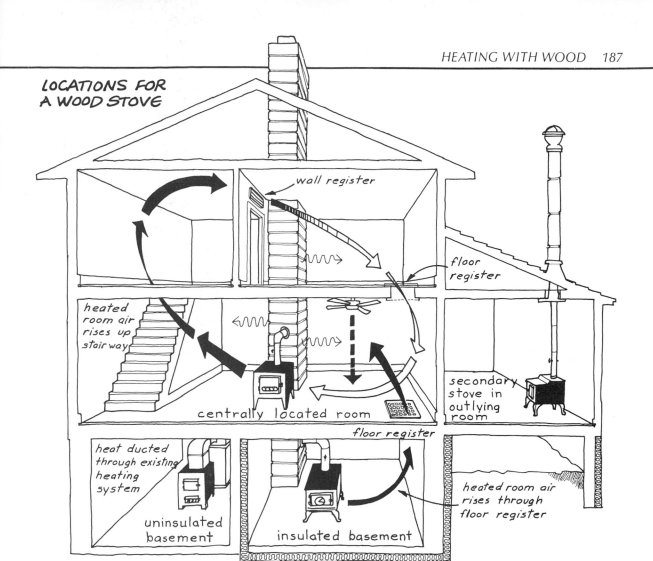

wall register

floor register

heated room air rises up stairway

centrally located room

secondary stove in outlying room

floor register

heat ducted through existing heating system

uninsulated basement

insulated basement

heated room air rises through floor register

It makes sense to put your wood stove in the most-used and most centrally located room in the house—usually the kitchen or living room/dining room area. Outlying rooms may need an additional stove or some way (fan or register) to transfer heat from a centrally located stove. If you have two floors to heat, place your stove by a stairwell or install floor and ceiling registers to take advantage of the fact that heat rises. You can also place your wood stove in the basement. If you have an uninsulated basement, there's no point in heating that space, so it's best to duct its heat upstairs, through your existing system. If, on the other hand, your basement is insulated, it might just be best to cut registers in the floor above the stove and let heated room air travel upstairs that way. An open staircase could take the place of floor registers.

WOODSTOVE-FIREPLACE CONNECTIONS

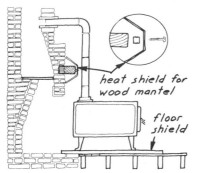

heat shield for wood mantel

floor shield

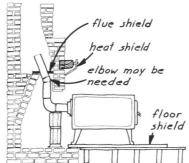

flue shield

heat shield

elbow may be needed

floor shield

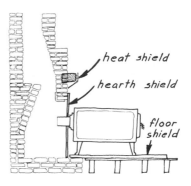

heat shield

hearth shield

floor shield

The energy-efficient thing you can do with your fireplace is to insert a wood stove into it. There are several ways to make the installation. A freestanding stove can be connected by a stovepipe that runs through the masonry over the mantel. If the mantel is wood, install a heat shield.

Another way to make a stove-fireplace connection is to run the stovepipe right into the fireplace flue. Remove the fireplace damper and replace it with a flue shield, or, even easier, remove the damper and cover the fireplace mouth with a hearth shield. Then vent the stove through the shield.

excessive amounts of creosote build up inside the pipe. Such cooling can also cause backdrafts of smoke into the room.

Wood for Your Stove

All wood contains water, which must be evaporated for the wood to burn properly. Green wood contains up to 50 percent moisture by weight, and this high water content makes igniting and burning it slow and difficult. This means low combustion efficiency. Dry wood, on the other hand, contains about 20 percent water, burns readily, and produces lots of heat.

Green wood is generally less expensive than seasoned wood, so you might be able to save money by buying it early and drying it out yourself. Wood should be stacked for at least six months to properly season it. To get it much drier, you'd have to heat it in a kiln, a luxury few wood stove owners can afford.

If you cut your own wood, do it in spring so that by winter it's had a six-month seasoning. If you buy wood, learn how to check its moisture content. Look at the end of a log for the darkening and cracking that indicate shrinking (drying) has taken place. Another test is to hit a couple of pieces of the wood together. A "thud" probably means wet wood; a sharp "crack" drier wood. Remember that seasoned wood will also be considerably lighter than the green variety.

Buying Wood

Hardwood is heavier than softwood and contains more Btus per pound; it also costs more. Softwood provides less heat and takes up more space in the woodpile for a season's supply, but it's cheaper. Table 5-3 ranks various kinds of wood.

UNITS OF WOOD

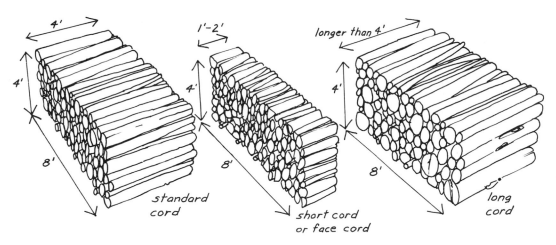

standard cord

short cord or face cord

long cord

Steer clear of punky, rotten wood, even at a bargain price. And be careful about resinous pitch pine that burns like a torch for a short period of time and can make dangerously hot fires. The only thing it's good for is kindling.

Wood Stove Maintenance

Wood stoves need regular attention. Most days that just means unloading ashes and loading wood. But once a year—after the stove's been shut down for the season—it'll need some special attention. Here's a checklist of maintenance items:

Firebox

    Check for rust or burnout

    Check for burnt-out firebricks or replaceable firebox plates

    Clean out creosote

Doors

    Check for tightness (see page 181)

    Check gaskets

(Continued on page 192)

Firewood is usually measured by the **standard cord,** *which is a 128-cubic-foot pile of stacked wood. The wood is 4 feet long and is stacked 4 feet high and 8 feet wide.*

Don't be fooled into buying a **short cord** *(or face cord) at standard cord prices. Short cords are stacked in the same 8-foot-by-4-foot dimensions, but the wood is only 1 to 2 feet long.*

You're getting a good deal if you're sold a **long cord** *instead of a standard cord. Again, the stack is 8 by 4, but the wood is longer than 4 feet.*

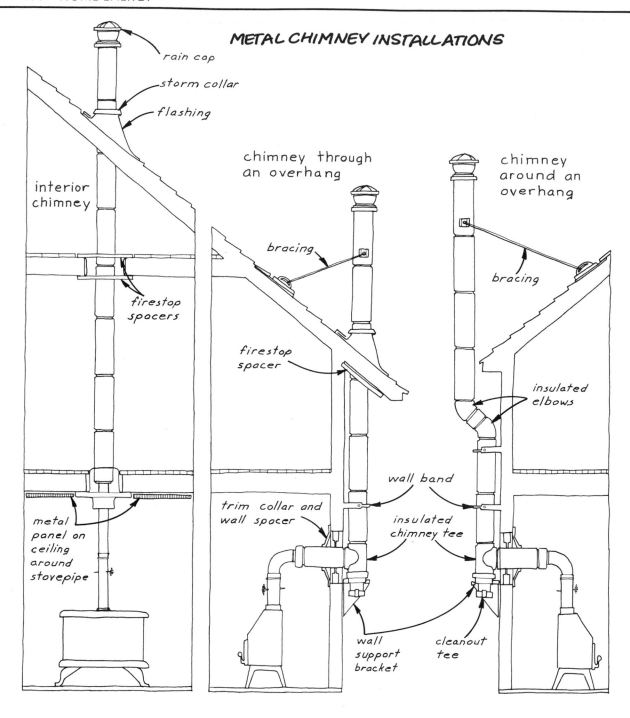

METAL CHIMNEY INSTALLATIONS

rain cap

storm collar

flashing

interior chimney

chimney through an overhang

chimney around an overhang

bracing

bracing

firestop spacers

firestop spacer

insulated elbows

metal panel on ceiling around stovepipe

trim collar and wall spacer

wall band

insulated chimney tee

wall support bracket

cleanout tee

If you don't plan to connect your wood stove to an existing fireplace, you'll have to install your own chimney. Metal ones are popular because they are inexpensive and come in ready-to-install kits. They can be installed inside or along the outside of your house.

Interior chimneys offer the advantage of heating up the rooms they run through. If you've got limited interior space, however, you can run the chimney up the outside of the house. Keep in mind that an outside chimney results in some heat loss and a chance of increased creosote buildup because flue gases are cooled more quickly by outside air. Here are three chimney installation options:

An interior chimney runs up from the stove through an insulated **through-box installed in the ceiling.** *The stovepipe can get pretty hot, so all of it should be 18 inches or more from all flammable surfaces. Shown in this illustration is a fireproof sheet of asbestos or metal panel on the ceiling around the pipe.*

When installing an exterior chimney, you can run the piping **through the overhang,** *or if the overhang is not wide enough, around it. Go through the overhang, if you can, because the flue should be as straight as possible to be the most efficient. If you must go* **around the overhang,** *be sure the elbows have extra clearance and are well insulated, because the slower-moving air at elbows causes them to get hotter than straight pipe runs.*

The rooftop opening that your chimney runs up through should be well flashed, and there should be a storm collar around the pipe. The clearance for a metal chimney is the same as that for a masonry chimney (see drawing on page 194).

Table 5-3
Firewood Ratings

Wood	Heat	Burns	Splits	Smoke	Sparks	Remarks
		Hardwoods				
Ash, beech, birch, dogwood, hard maple, hickory, oak	high	easy	easy	light	no	excellent
Cherry, soft maple, walnut	med	easy	easy	light	no	good
Elm, gum, sycamore	med	med	hard	med	no	fair
Aspen, basswood, cottonwood, yellow poplar	low	easy	easy	med	no	fair
		Softwoods				
Douglas fir, southern yellow pine	high	easy	easy	heavy	no	good
Eastern red cedar, white cedar	med	easy	easy	med	no	good
Cypress, redwood	med	easy	easy	med	no	fair
Larch, tamarack	med	easy	easy	med	no	fair
Balsam fir, eastern white pine, hemlock, red pine	low	med	easy	med	yes	fair
Spruce	low	easy	easy	heavy	yes	poor

SOURCE: John Vivian, *The New, Improved Wood Heat* (Emmaus, Pa.: Rodale Press, 1978).

MINIMUM STOVE CLEARANCES

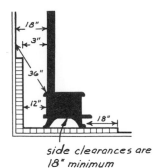

side clearances are 18" minimum

A wood stove can radiate a lot of heat, so it's important to install it at a safe distance from your wall. You also must protect your floors. Safe clearances will vary according to the type of fireproofing materials that you use. This illustration shows the minimum clearances needed for safe wood stove operation. Be sure to check with your stove distributor for specifics.

Seams

Make sure seams on cast-iron stoves are tight; recement if necessary

If welds on steel stove are broken, have them repaired

Paint

Touch up rusted or chipped areas with high-temperature paint or stove polish

Stovepipe/Flue

Clean out creosote to prevent a chimney fire

(Continued on page 194)

It doesn't take a lot of know-how to be your own chimney sweep, but it does take tools that are right for your particular chimney. There are two common chimney types: a vertical through-the-roof stack, and an exterior one with a cleanout tee.

To clean a through-the-roof chimney, you'll need a steel flue brush with handle extensions, gloves, a flashlight, bags, tape, and a drop cloth.

Lay down a drop cloth inside by the stove, put on your gloves, and disconnect the stovepipe from the stove and take it outside to clean later. Place a double-thick paper bag or heavy plastic garbage bag over the stovepipe opening to catch debris. Secure the bag with tape. You may want someone to stay inside to make sure the bag isn't dislodged. Now go up on the roof and insert the flue brush into the chimney. (A weighted brush makes the job easier.) After you've moved the brush up and down for a while, remove it and use the flashlight to check the chimney walls for creosote.

Another method for cleaning vertical through-the-roof chimneys and for chimneys with bends or turns involves two people and the use of a flue brush with ropes. Again, remove the stovepipe and place a bag over the opening, but cut a small slit near the top of the bag. Secure the flue brush with ropes at either end. Go up on the roof and lower the bottom rope to the person in the house, who should carefully pull the rope through the slit in the bag and then pull the brush to the bottom of the chimney. Then slowly pull the brush back up to the top of the stack.

Cleaning an exterior chimney with a cleanout tee is relatively easy and requires a flue brush with extensions. Remove the stovepipe and clean it outside. Place a bag with a slit in it over the opening. Attach another (uncut) bag to the end of the opened cleanout tee outside. Clean the horizontal pipe from the interior first. Leave the interior bag in place while you go outside and up on the roof to clean the vertical section. You can also remove the outside bag and clean the vertical section through the cleanout tee.

CLEANING A CHIMNEY

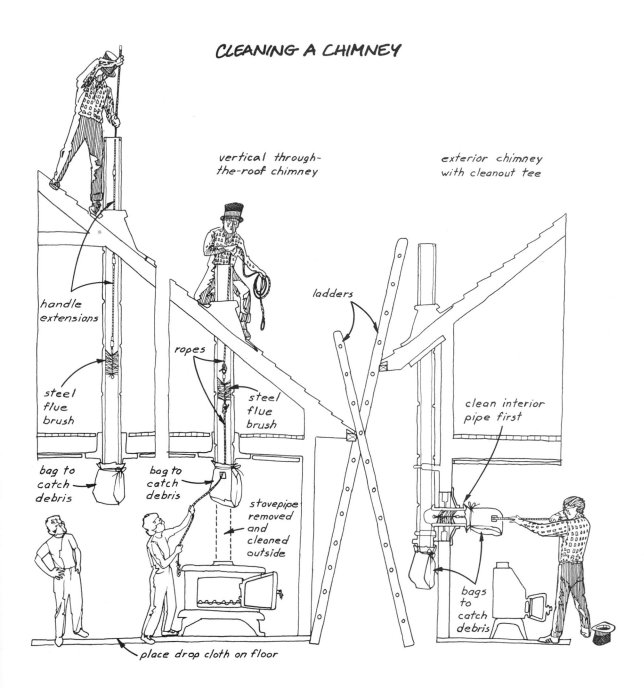

vertical through-the-roof chimney

exterior chimney with cleanout tee

handle extensions

steel flue brush

bag to catch debris

ropes

steel flue brush

ladders

clean interior pipe first

bag to catch debris

stovepipe removed and cleaned outside

bags to catch debris

place drop cloth on floor

Chimney Cleaning

Creosote isn't much of a problem inside the stove, because it generally stays hot enough in there to prevent much creosote from building up. But you'd better stoke up a good hot fire now and then and leave the air inlets open for 15 minutes or so to burn the tarry gunk out of the stovepipe and flue to prevent a chimney fire. Once or twice a year clean out the stovepipe and chimney.

Fireplaces

According to the Department of Energy, the efficiencies of most fireplaces range from +10 to −5 percent. This documents the worst suspicions of those who believe fireplaces can actually be energy losers because they draw heated air out of the room and send it up the chimney, in turn sucking more cold air into the house to replace the warm air that went up the chimney.

Improving Fireplace Design

A good fireplace has the fire well out into the room, not tucked back in the chimney. It has a high lintel to let lots of radiant heat into the room, and much more thermal mass than a poor fireplace has. Even if your fireplace flunks the design test on all points, however, there may still be some things you can do to improve its performance.

All fireplaces smoke, but some of them send smoke into the room instead of up the chimney. This could happen because kitchen or bathroom vents are exhausting air the chimney needs, or your house is so tight that there just isn't enough fresh air to feed the fireplace properly.

If the latter is your problem, you can bring in more air from the basement, or even from outside by opening a window a crack. If you provide an outside cold-air supply that feeds directly to the fireplace, warm air from inside the house won't be lost.

The problem could also be in the chimney itself. You must be sure the chimney is at the proper clearance. This applies to wood stoves as well as fireplaces.

If there's a damper in your chimney, be sure it's opened the proper amount. And if there isn't a damper, you'd better see about having one put in. You wouldn't tolerate a hole the size of your chimney in one of the walls, and air goes up the chimney faster than through a hole in the wall.

Check also to see if there's a smoke shelf in your fireplace. The purpose of the smoke shelf is to curl descending chimney air around

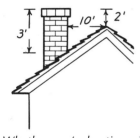

CHIMNEY CLEARANCES

Whether you're heating with a fireplace or wood stove, your chimney must be installed with the proper clearance. The top of the chimney should be at least 2 feet higher than any roof section or anything within 10 feet of it, and at least 3 feet higher than the roof at the point where the chimney emerges.

A freestanding fireplace is easy to install and can be put anywhere that a stovepipe can be run through the ceiling or wall. Know, though, that you trade the thermal mass and the good draft control of a wood stove for that 360° view of the fire within.

and back up into the rising warm air, solving the smoke problem. There may also be obstructions in the chimney, like bird or squirrel nests!

Freestanding Fireplaces

A freestanding fireplace isn't tied to a wall like the conventional fireplace, and these "put-them-anywhere" metal fireplaces have many advantages over conventional masonry. Installation is easy because they can be put just about anywhere that a stovepipe can be run to

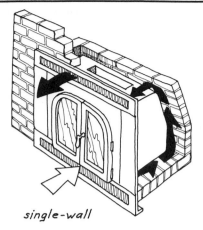

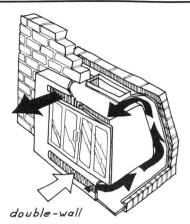

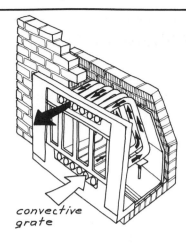

single-wall

double-wall

convective grate

FIREPLACE INSERTS

If you have a heat-gulping fireplace, but don't want to retrofit it with a free-standing wood stove, you can buy a fireplace insert to boost efficiency. Inserts aren't as efficient as wood stoves are, but they are less obtrusive, and that can mean a lot if your fireplace is in a room with little floor space to spare.

Fireplace inserts come in a variety of styles, but they share many of the same features. All inserts have two sets of air intakes—one for convective air and one for combustion air. The two air streams are kept separate, so no smoke or combustion gases mix with the air flowing into the room. Practically all inserts have electric blowers to help circulate the convective air.

There are three basic types of fireplace inserts: single-wall, double-wall, and convective grate. In a **single-wall** design, room air enters at the bottom of the unit, and rises by convection between the back of the insert and the rear wall of the fireplace. It returns to the room through a grate at the top of the unit.

Double-wall inserts are the most commonly found. Again, room air enters through a bottom grate, but the convective air circulates between the insert's two walls instead of between the insert and fireplace wall.

A **convective grate** provides heat by circulating room air through C-shaped tubes that act as a grate.

the ceiling or wall. In fact, installing a freestander is much like installing a stove. A protective hearth is needed, similar to that required for a stove, and proper clearances must be maintained from walls and floor.

Circulating Fireplaces

Circulating fireplaces are exceptions to the general rule that fireplaces are inefficient. Air circulators are the most effective of all fireplaces, and some experts say that the best of them can match the performance of a so-so stove.

Fireplace Doors

In theory, fireplace doors restrict the airflow into a fireplace, thus cutting heat losses. Supposedly, the fire receives only small streams of air coming through small vents in the doors.

The theory is great. But in practice, many glass doors aren't airtight enough to work as advertised. So look for a set with well-sealed, close-fitting doors. And be sure you can seal the set's outer frame against the face of your fireplace. Some companies, for instance, make sets of doors having clamps that attach to the fireplace lintel, and brackets that can be bolted to the fireplace floor. But even clamps and brackets won't necessarily seal all gaps between the outer frame and either the fireplace face or the hearth, especially if you have a stone fireplace with uneven surfaces. You'll need to plug the gaps with refractory cement, ceramic wool, or fiberglass that has no foil or paper backing.

Experts disagree about the best way to use fireplace doors. Because glass and metal block radiant heat, some authorities say you'd be smart to leave the doors open when the fire is at its peak. You can keep sparks off your floor by closable fireplace screens that are available with most door sets. You should receive more than enough heat to compensate for the warm room air traveling through the screens and up the flue. Then, as the fire is dying—when the fireplace is contributing less heat than it's wasting—close the doors.

But there's also another school of thought. According to scientists at Lawrence Berkeley Laboratory in California, you can leave fireplace doors closed even when the fire is hottest, provided you have tight-fitting doors that really stem the flow of air toward the fire. The fireplace will then be a sealed unit, heat will be trapped inside, and the firebox temperatures will rise higher than in an open fireplace. The doors will intercept much of this heat, but enough will reach the room to make things comfortable, and you will have tamed the fire's craving for room air. On balance, you'll come out ahead.

A third possibility is to install fireplace doors and a convective grate. Many of these grates are designed to be compatible with doors, and the two accessories make a good team. Keep the doors closed to block room air from the fire; rely on the grate for heat. (Before deciding to keep glass doors closed throughout the burn—whether or not you're using a convective grate—be sure the doors were designed to withstand the high temperatures in the firebox. Otherwise, the glass could shatter.

There are a variety of designs for circulating fireplaces, but they all function pretty much the same. They radiate heat from their walls just like traditional fireplaces, but unlike conventional fireplaces, they're designed to heat room air in an interior chamber (either masonry or metal). The air heated in this chamber rises and circulates into the room through registers at the top of the fireplace, while cooler air from the room enters the chamber through openings at the bottom of the fireplace.

Fireplace grates made from hollow tubes bent into the shape of a C can increase fireplace efficiency by circulating air heated by the fire into the room. Some designs use automobile exhaust pipe, which won't hold up under the intense heat of a good fire; if you get a circulating grate, make sure it's made of heavy-gauge metal.

Fireplace Inserts

Installing a wood stove in your fireplace is a good way to deal with its inefficiency. But many people haven't the space for such an installation or for other reasons don't want to use a stove. In such cases, the simplest solution is to add a good fireplace insert.

Here's how a fireplace insert works. Cool air near the floor is circulated around the firebox and back into the room. Most of the new generation of fireplace inserts have blowers, which increase air circulation but are often noisy and, of course, consume some electricity. Hear one in action before buying it.

The main problem with these inserts is creosote. Its buildup is rapid and heavy, especially on the smoke shelf of the chimney and also in the firebox. Catalytic combustors (see page 169) are now available for fireplace inserts; in addition to making the inserts heat more effectively, they should also cut down on the creosote problem.

Inconvenience is another problem with inserts. They weigh as much as 600 pounds, and removing one from the fireplace for cleaning is a difficult and messy job. Some manufacturers solve this problem by putting wheels on the bottom of the insert for easy rolling.

Central Heating

Wood and coal furnaces and boilers have also made a bit of a comeback with the wood stove revival. Cleaner and more efficient than they used to be, some can use oil, gas, or electricity as well. This makes it possible to switch to conventional fuel when there's no one home to load in wood or coal.

Central wood heat can be used with gravity hot-air, steam heat, or circulating hot-water systems. Prices range from $500 to $1,000, for a small add-on wood heater, to thousands of dollars for a full-scale

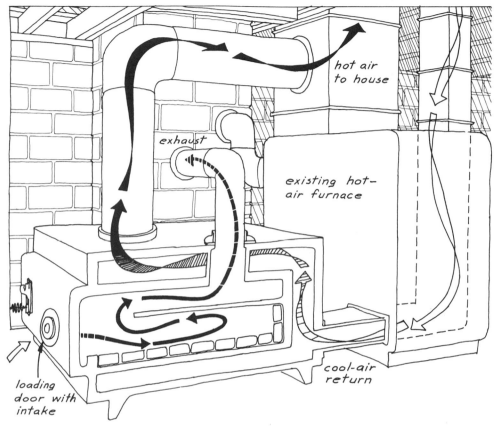

hot air
to house

exhaust

existing hot-
air furnace

loading
door with
intake

cool-air
return

A WOOD-FIRED CENTRAL HEATING SYSTEM

If you want central wood heating and you've got a large house, a wood furnace
might be the way to go. And if you tie the wood furnace into your existing
heating system, you can keep your original furnace in place as a backup
heating unit. A wood furnace itself is simply a wood stove enclosed in a metal
jacket that collects heat and distributes it to the house through ducts.

Heating with a wood furnace has some drawbacks. A wood furnace could
cost twice as much as an oil or gas furnace, and if your house is small, you may
be able to adequately heat it simply with two wood stoves located in the
living areas.

You may have to build a new chimney for a wood furnace. They require
Class A chimneys—lined masonry or factory-built all-fuel types. A wood
furnace and its chimney need frequent cleaning, sometimes as often as once
a month, to ensure against heavy creosote.

It goes without saying that you need to have an abundant supply of wood
to keep your hungry furnace going throughout the heating season. And check
your local building codes—some don't allow the use of wood furnaces.

wood furnace. The latter can deliver up to 1.2 million Btu an hour, about 10 times the output of a large residential wood stove. A few firms still manufacture these wood central-heating furnaces.

A cheaper approach is to add a large wood stove designed for use with an existing fossil-fuel furnace. These are available for hot-air or hot-water systems and are designed to connect to existing automatic heating systems.

In most cases, however, a wood stove upstairs will probably be a better solution than a wood central-heating plant in the basement. A wood furnace means lots of heavy wood going down and lots of messy ashes coming up.

For More Information on Heating with Wood

Homeowner's Guide to Wood Stoves, by the editors of Sunset Books, 1979: Sunset Books, Lane Publishing, Menlo Park, CA 94025

Just about all you need to know about wood stoves, plus beautiful color photos of dozens of installations of every kind.

The New, Improved Wood Heat, John Vivian, 1978: Rodale Press, 33 E. Minor St., Emmaus, PA 18049

This updated edition of Vivian's earlier *Wood Heat* has to be the best book available on every aspect of heating with wood. Stoves, fireplaces, chimneys, cooking, maintenance, and wood itself are all covered completely and entertainingly by an expert.

Suppliers

Prices and addresses given are as of this writing; check with the supplier before sending money.

Fireplace and Stove Plans and Kits
Albie Barden (RD 1, Box 640, Norridgewock, ME 04957) offers plans for an end-loading horizontal-run stove and for a center-loading Finnish-style stove. The plans cost $12 for the end-loader, $15 for the Finnish.

Hahsa (Box 112-S, Falls, PA 18615) offers plans for Hahsa, "the safe outdoor wood furnace." This wood burner is built some distance from the house for safety. Underground ducts carry the heat to the house. The plans cost $20. An information booklet costs $1.

Basilio Lepuschenko (RFD 1, Box 589, Richmond, ME 04357) offers a 30-page booklet with complete plans and instructions for building and using a large masonry stove based on the stoves the author remembers from his childhood on the Baltic coast. Plans and information are given for building three-, five-, and seven-flue masonry stoves. The booklet costs $11.

Locke Stove Company (114 W. 11th St., Kansas City, MO 64105) offers a complete set of castings to make your own barrel stove, including set of legs, front-feed door, and flue outlet collar. Write for the current price.

Radiant Mass Fireplaces (Box 1637, Idaho Springs, CO 80452) offers construction plans for the Radiant Mass Fireplace, a 3-foot-by-7-foot-by-8-foot site-built masonry fireplace with inner baffled masonry flue chambers that deliver heat to the outer brick surface at 90°F to 125°F for 16 to 20 hours per firing. Its efficiency exceeds 85 percent, and there is no creosote buildup, and minimal pollution. Radiant Mass Fireplaces builds the fireplace complete in the Denver/Front Range area for $5,000. An owner capable of doing the masonry might do the job for about $3,000. The plans cost $65. A hardware kit containing cast-iron airtight doors with high-temperature glass, air intake and exhaust dampers, arch template, inspection/cleanout doors, and insulation is available for $650.

CHAPTER 6

Energy from the Wind

Wind power is an age-old form of power that brings to mind tall sailing ships, picturesque Dutch windmills, and early American homestead water pumpers. Modern engineering may have stolen some of the romance of the wind, but it has given us in its place wind generators that make it practical, in select places, to use the wind as a major source of electricity.

Remote sites off the utility grid are more attractive for wind machines because of the great expense of running in a power line, and also because there are fewer legal and other restrictions in remote areas. On the other hand, a sizable number of modern wind machines have been installed in cities and interconnected with utility grids as a convenient backup source of electric power.

A recent Department of Energy survey of the performance of residential wind machines showed that average performance was low. But for people with a passion for whirling blades and the feel of tapping the natural energy of the wind, a wind machine can be a source of great satisfaction even when the economics are marginal. And with careful evaluation of the local wind resource and intelligent installation of the proper size and design, a wind machine can be cost-effective.

Introduction to Wind Power

When the current surge of interest in wind power began around 1970, an estimated 200,000 of the oldtime windmills were still working. Out of the renewed interest came a new generation of electricity-

What You'll Learn in This Chapter

- The many legal considerations to installing a wind machine (page 208)

- Wind machines are potentially hazardous devices, and safeguards must be provided to assure proper operation and safety for you and those living near you (page 210)

- Unless your site has average winds of 10 mph or more, and unobstructed access to that resource, don't invest in wind power (page 211)

- How to correctly size a wind machine to fill your needs and to be economical (page 216)

- How to connect to the utility grid for backup power and sell the power company any surplus wind electric power you produce (page 216)

- A stand-alone wind power system requires a battery storage system for continuous operation (page 221)

- Wind machines require regular maintenance; and so does the stand-alone battery storage system (page 223)

- Wind machine options include ready-built systems, rebuilt equipment, kits, and plans (page 225)

- Wind machines cost $1,500 a kilowatt and up, and most homes use from 3 to 5 kilowatts of electric power (page 225)

producing wind machines. Hundreds of large wind machines are now feeding megawatts into utility grids that serve industry and home-owners, and thousands of small wind machines provide direct power to homes and buildings, some of them also feeding power into utility grids.

Is Wind Power for You?

The Department of Energy booklet *Is the Wind a Practical Source of Energy for You?* begins with a checklist for answering that question. The list is a very helpful starting point:

- Evaluate potential legal and environmental problems
- Evaluate your energy requirements (if your electrical needs are high, wind probably won't be able to meet them all)
- Evaluate the wind resource at your proposed location
- Evaluate your application (is wind the best way for you to meet your electrical needs?)
- Select system and components

The past meets the present: a 50-year-old windmill is still pumping water, while its new neighbor generates power for the residents of Clayton, New Mexico.

- Evaluate the cost of the system
- Evaluate alternatives in buying, installing, and owning a wind system

Economics

At present, small wind-machine electric power costs 15 cents a kilowatt-hour (kwh) and more. However, the industry believes that this can be cut to as low as 5 cents a kilowatt-hour with improved designs and mass production. In fact, a large wind machine being built by General Electric for installation in Hawaii is expected to produce power for 4 cents a kilowatt-hour. And once a wind machine is installed, the price of the electricity it generates won't change.

If you live in Seattle, Washington, or Kalispell, Montana, or somewhere else where hydroelectric power costs only 2 cents a kilowatt-hour, a wind machine hasn't a prayer of being cost-effective. Putting up a wind machine would be like installing a residential solar heating system in a national forest, with all the free firewood you need!

To get a sense of the economics of a wind generator for your situation, we'll estimate how large a wind machine you'll need to take care of your present electric power load. The simplest way to do this is to find the number of kilowatt-hours you use, and the best place to look is your utility bill.

If you use energy sparingly, and have put Chapter 1 to good use, you may be one of the fortunate rate payers using less than 500 kwh a month. Electric costs also depend on whether you pay 2 cents or 10 cents a kilowatt-hour. Let's see how the cost per kilowatt-hour affects a system's cost-effectiveness. As we do the math, keep in mind

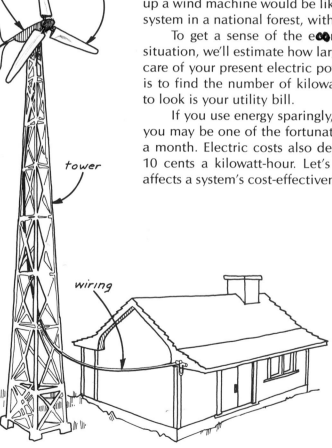

generator

rotor

tower

wiring

BASIC COMPONENTS OF A WIND SYSTEM

Here's a general view of basic wind system components. This is just one example of the many rotors, generators, and towers available.

*The **rotor** converts the wind's energy into rotary motion. The **generator** then converts that rotary motion into electric current. The **tower** supports the rotor/generator assembly at the proper height and location, and it houses the **wiring** that is hooked up to the house's electric gear, either stand-alone or utility-connected.*

The height of a wind tower depends upon local ground turbulence in the path of prevailing winds. One can imagine what it took to erect this one!

that it doesn't take into account any increases in electric power costs, and that a 7 percent annual cost increase will quadruple utility rates in 20 years.

$$500 \text{ kwh @ 2 cents} = \$10 \text{ a month}$$
$$= \$120 \text{ a year}$$
$$= \$2,400 \text{ in 20 years}$$

To compete with 2-cent electricity, you'll have to buy, install, and maintain a wind machine for 20 years at a total cost of $2,400 to break even at the present cost of your electric power.

Now let's try 10 cents a kilowatt-hour:

$$500 \text{ kwh } @ 10 \text{ cents} = \$50 \text{ a month}$$
$$= \$600 \text{ a year}$$
$$= \$12,000 \text{ in } 20 \text{ years}$$

As you can quickly see, at 10 cents a kilowatt-hour, you have a lot more money to play with.

A demand for 500 kwh a month calls for a wind machine rated at about 5 kilowatts of power in a 12-mph wind. The cost for such a machine is about $7,500. Add about $2,500 for tower, installation, and miscellaneous, plus $2,500 for maintenance and insurance for 20 years:

$$
\begin{array}{r}
\$\ 7,500 \\
2,500 \\
\underline{2,500} \\
\$12,500
\end{array}
$$

Even with no increase in utility charges for electric power, and no federal or state tax credits, this system comes close to breaking even in 20 years if you're paying 10 cents for each kilowatt-hour. When the 40 percent federal tax credit and a 20 percent state tax credit are deducted (see Appendix C), the wind system cost drops to $5,000. Depending on the interest charges, the system would pay for itself in 10 years or so and save about 50 percent on electric power over its· 20-year lifetime.

You can plug your own numbers into the equation above. If you use table 6-1 to find the monthly kilowatt-hour output of the wind system you're planning, simply multiply that number times the cost of utility electric power times 12 (months) times 20 (years) to get the 20-year dollar value of the system's energy savings. As in the above examples, you can compare this number with the cost of buying, installing, maintaining, and insuring the system for 20 years, a figure you should be able to get from your wind system sales representative.

The new wind machines don't yet have a long track record, but considering that some ancient Jacobs and other pioneer wind generators are still operating, 20 years may be a very conservative estimate of service life, in which case the above-mentioned installation would be that much more profitable.

Rules, Regulations, and Precautions

If you plan to put up a wind machine in a remote location, you'll have a minimum of legal considerations to worry about. The closer you get to downtown, however, the more such problems are likely. And if you plan a grid-connected system, you must abide by utility regulations. Other things you must consider are codes, zoning, wind rights, safety, and insurance.

Codes

If your home is in a heavily populated urban area, you'll have to comply with building codes. Remember that a wind machine must be mounted on a tower a considerable distance above the ground. Your neighbors won't be the only ones concerned with such a project; local officials will also have a great deal to say. Generally, you'll have to obtain a zoning variance to install a tower in a densely populated area.

Any high tower must be properly designed, built, and rigged. A wind machine has the added problem of supporting a heavy generator and a large, rotating propeller. Such a device may also be considered an "attractive nuisance," and you may be required to put up gates and fences to keep people away from it. In addition to building codes, local electrical codes also apply. This is true even if you don't intend to connect to the utility.

Zoning

Most communities of 5,000 or more have zoning ordinances; all states have enabling legislation for such ordinances. Zoning basically divides an area into residential, commercial, and industrial zones. Putting up a wind machine in an industrial area should be little problem. But commercial and residential areas are a different story. Even though there may be no specific bans of wind machines as such, other things like height regulations and property line setbacks may rule them out. Zoning also includes aesthetic considerations, and some people in your area may consider windmills unsightly.

It's possible to request and receive zoning variances. Success will depend largely on the surrounding community and its attitudes toward wind machines. If there's a wind machine not far from where you hope to install one, you have an advantage. Talk with the owner for details. He succeeded; maybe he can help you do the same.

In rare cases, TV interference may be a problem. Caused principally by large metal blades, its more likely to affect the upper UHF channels. Under certain conditions, interference can extend for several miles. Fiberglass blades aren't likely to cause interference.

Wind Rights

In early times, villages banned the growing of trees that could block windmills, but the situation today is usually much different. Only California, Florida, Nebraska, and Texas mention wind machines in their statutes. However, the state tax credits for wind machines are encouraging. Having passed such laws, it would seem wise for governments to protect an owner's investment with wind rights statutes. The sun rights problem is often handled by covenants or easements; this may be a model for assuring access to the wind.

Safety

Most wind machines are very safe, but it would be negligent not to discuss some problems that have cropped up in recent wind power history. Some home-built machines have had serious failure of blades and/or towers. Be sure that your wind machine has effective overspeed protection in the event of high winds and that your tower is properly installed and maintained.

If you're subject to thunderstorms, better provide lightning protection for your wind machine. Lightning seeks projections above the terrain; metal projections are that much more attractive. Many old water-pumping windmills have been left standing for that very reason —to serve as lightning rods to attract lightning away from a ranch-house or other building in the area. A real lightning rod, properly grounded, offers protection for your house and for the wind machine itself. Lightning generally damages exposed electric wires, not a grounded tower, so bury the wiring.

Insurance

Insure your wind machine—it represents a sizable financial outlay. You must also have coverage for personal liability and property damage. Situations differ, and so do insurance companies and their agents. If there are a dozen wind machines already operating in the neighborhood, you shouldn't have trouble getting proper insurance coverage. But if you're the first one on the block with wind power, you may have some difficulty.

The Essential Ingredient: Wind Speed

There are probably a dozen or so prime considerations when planning a wind system, but three in particular stand out:

The speed of the wind
The size of the wind machine required
The cost of such a machine

Most people overestimate the first consideration and then apparently try to make up for that by greatly underestimating the others. Yet wind speed is the essential ingredient in using wind power. Here's the first and most important wind RULE OF THUMB:

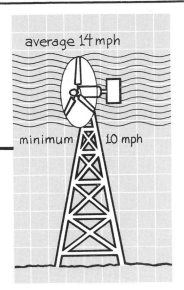

The wind must blow regularly and strongly to be cost-effective. An average year-round wind speed of about 14 miles an hour is desirable; if it's not at least 10 mph, better not consider investing in a wind machine!

Measuring the wind accurately isn't easy; a wet finger won't do it. A scientist at Oregon State University checks the wind with a 50-cent kite and a small spring scale, but he's an exception. Unless you know for a positive fact that you have plenty of wind at your site, carefully read the next two sections before going on to cost and design considerations. The finest wind machine you can buy will be of little use if you don't have the wind to keep it spinning a good bit of the time.

A Wind Primer

Below 10 mph, there isn't enough energy in the wind to do much good. But winds above 30 mph can exceed the strength of propeller blades. So most wind machines are designed to turn 90 degrees from the wind, brake, feather, or otherwise protect themselves from damage by strong winds.

To understand why wind speed is so critical to the success of a wind machine, consider this RULE OF THUMB:

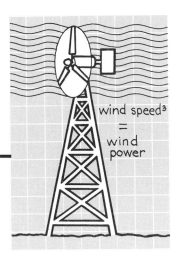

Wind power is proportional to the cube of wind speed.

Here's the cube rule in action:

$$\text{The cube of } 10 = 1,000$$
$$\text{The cube of } 8 = 512$$

This means that a wind machine that delivers 1,000 units of power in a 10-mph wind delivers only 512 units in an 8-mph wind. That's a power difference of almost 50 percent with the wind only 2 mph slower.

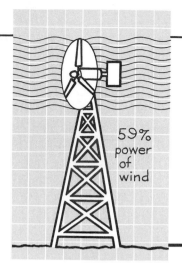

59%
power
of
wind

But the cube rule works the other way—to your advantage—if average wind speed is faster than you assumed. For example, if wind speed is twice what you thought, you'll have eight times as much power:

The cube of 10 = 1,000
The cube of 20 = 8,000

The cube rule was worked out in the 1920s by Alfred Betz. From his work came another wind RULE OF THUMB:

A perfect wind machine would extract about 59 percent of the power in the wind.

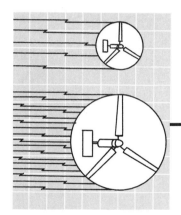

There are no perfect wind machines, of course, and an overall efficiency of about 35 percent is very good performance. Here's another important RULE OF THUMB:

Wind power is proportional to the area swept by the propeller or rotor of the wind machine.

Thus, for more power at a given wind speed, larger blades are needed. Area is about 3.14 times the square of the radius of the propeller (which is the length of the blades). So a propeller with 3-foot blades sweeps an area of about 28 square feet (3.14 × 3 × 3). Six-foot blades sweep about 113 square feet. Notice that four times as much wind power is produced from a blade twice as long.

Evaluating Your Site

If you're building a house and would like to include wind power, it'll pay you to do some wind prospecting before you purchase a lot. If you want to use wind power on your present lot, the task is simpler: you either have enough wind at your present site or you don't.

Wind speed varies with the time of day and with the seasons. It varies with place, too. The Texas Panhandle and parts of the Pacific Coast have strong winds most of the time. So do parts of New England. But winds in many other parts of the country aren't as dependable. The wind map here will give you an idea of wind strength in your region.

Portland, OR (9.1)
Helena, MT (9.7)
Bismarck, ND (12.7)
Montpelier, VT (9.6)
Bangor, ME (9.5)
Boise, ID (10.4)
Rapid City, SD (12.18)
Minneapolis, MN (12.5)
Boston, MA (15.7)
Buffalo, NY (15.3)
New York, NY (14.5)
Reno, NV (7.8)
Salt Lake City, UT (10.3)
Denver, CO (11.7)
Chicago, IL (12.9)
Kansas City, MO (12.5)
Washington, DC (11.5)
San Francisco, CA (12.7)
St. Louis, MO (10.5)
Santa Fe, NM (13.7)
Nashville, TN (9.9)
Los Angeles, CA (7.9)
Amarillo, TX (15.6)
Atlanta, GA (11.3)
Phoenix, AZ (6.4)
Dallas, TX (12.1)
San Antonio, TX (10.8)
New Orleans, LA (10.7)
Miami, FL (10.4)

WHERE THE WIND IS

Airports are usually the handiest sources of wind information, but unless you live close to an airport, its records won't tell you about the winds in your own backyard. Keep in mind, too, that airports generally aren't built in areas of known strong winds.

If the airport records are encouraging, it's time to start a serious monitoring program at your site. A recording anemometer, or wind odometer, is best because you should have a record of actual wind performance at the site where you plan to install a wind machine. An odometer records the average wind speed, which is what you must know to accurately estimate the wind potential in your area. Don't get a simple anemometer of the type used for home weather stations. These indicate only the instantaneous wind speed; average wind is what you must know.

Ideally, you should record an entire year of average wind speed—and direction, too, if possible. Several months is minimum. Daily readings are better than weekly or monthly. Stronger winter winds are typical.

Although factors such as trees, hills, or houses determine how much wind is available at any particular site, some regions are windier than others because of topographic features like mountains and plateaus. The numbers on the map represent approximate average annual wind speeds in miles per hour for the specific cities shown and have been calculated to represent the winds likely to occur at the top of a 90-foot wind machine tower.

Kilowatts in the Wind

cabin-size
(6'-12')

home-size
(12'-16')

all-electric-size
(16'-40')

Table 6-1
Estimated Energy in Kilowatt-Hours per Month

Rotor Diameter (ft)	Average Wind Speed (mph)*				
	8	10	12	14	16
2	3.4	6.5	11.3	17.9	26.7
4	13.4	26.0	45.0	71.6	106.8
6	30.0	58.7	101.4	161.0	240.2
8	53.4	104.3	180.2	286.2	427.0
10	83.4	163.0	281.6	447.1	667.1
12	120.2	234.7	405.5	643.8	960.6
14	163.5	319.4	551.9	876.3	1,307.5
16	213.6	417.1	720.8	1,144.6	1,707.8
18	270.3	527.9	912.2	1,448.6	2,161.4
20	333.7	652.8	1,126.2	1,788.3	2,668.4
22	403.8	788.6	1,362.7	2,163.9	3,228.8
24	480.5	938.5	1,621.8	2,575.2	3,842.5
26	564.0	1,101.5	1,903.3	3,022.3	4,509.6
28	654.1	1,277.4	2,207.4	3,505.1	5,230.0
30	750.8	1,466.4	2,534.0	4,023.8	6,003.9
32	854.3	1,668.5	2,883.1	4,578.1	6,831.0
34	964.4	1,883.5	3,254.7	5,168.3	7,711.6
36	1,081.2	2,111.6	3,648.9	5,794.2	8,645.5
38	1,204.6	2,352.8	4,065.6	6,455.9	9,632.8
40	1,334.8	2,607.0	4,504.8	7,153.3	10,673.5

*Based on efficiency of 50% at the rotor (although by the time power gets from rotor, through the circuitry, and to the point of use, it drops to about 30% efficiency)

Don't measure the wind at ground level, but as close to the height of a tower-mounted wind machine as you can. You'll get a more realistic reading because wind velocity is less close to the ground because of surface friction. Here's a rough RULE OF THUMB:

Wind speed 30 feet above the surface is about 10 percent faster than at the surface.

If you're planning to do a wind site analysis to see if it's right for a wind electric system, you can use table 6-1 to find out what the data mean. The bottom line for feasibility is the amount of power a wind system will produce, and of course the more wind you have, the more kilowatt-hours you'll get.

This table can be used in a couple of ways: say, for example, that your site analysis tells what the average wind speed is. You can take that number and correlate it with the size of the wind machine you have in mind (based on rotor diameter) to find an estimate of monthly kilowatt-hour production. Even before a site analysis is done, you might already know how many kilowatt-hours you want per month. If you want a minimum of 500 kwh per month, you can find the minimum rotor size required per a given average wind speed. With an 8 mph average wind speed, you'd need at least a 26-foot rotor to get that much power, or an 18-foot rotor with a 10 mph average. Keep in mind that these are very basic estimates that are made from purely average wind speed. The averages at your site could vary widely through the year. If you have utility-supplied electricity at your site, you should compare its cost with the cost of wind-generated electricity. The value of the wind system output would simply be the monthly kilowatt-hour output times 12 months times the cost (in dollars/kwh of utility power). This amount would be reduced by the annual cost of maintenance, which could be 1 to 3 percent of the total system cost.

Wind power systems can be categorized in three size ranges. **Cabin-size** *systems, with 6- to 12-foot rotors, are meant for minimal energy-use situations, such as powering a few lights, a television, and some low-power appliances. When the average wind speed is 12 mph, cabin-size systems can produce 100 to 400 kilowatt-hours per month.*

Home-size *systems, with 12- to 16-foot rotors, are capable of powering all or most of your home's electric needs. With average wind speeds of 12 mph, home-size systems produce about 400 to 700 kwh a month.*

The **all-electric** *system, with 16- to 40-foot rotors, is the largest and most powerful, and also the most complicated. These systems can power an all-electric home. Most are designed for utility connection. All-electric systems produce 700 to 4,500 kwh per month when average wind speeds are 12 mph.*

This is very important, because a 10 percent increase in wind speed produces about 33 percent more power. Wind direction is important, too, unless your site is on a knoll above all obstructions. Wind turbulence over trees and other obstructions should be taken into account. Here's a siting RULE OF THUMB:

The center of the rotor should be at least 30 feet higher than any obstacle within 300 feet.

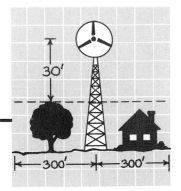

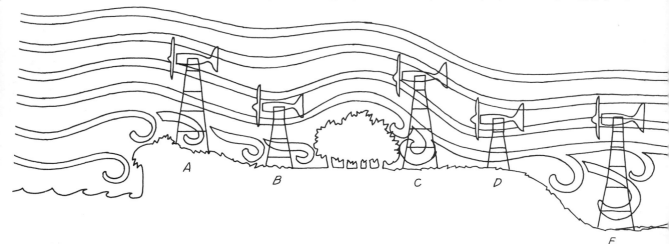

WIND TURBULENCE OVER DIFFERENT TERRAINS

While you can't actually see the airflow patterns over land, you can make some generalizations about air currents. Here are some of the basics.

Shore areas usually have good winds, but cliffs or dunes can create severe turbulence. Place your wind machine on a tall tower (A) to be above the turbulent air flow. A little farther inland, the turbulence lessens, and you can use a shorter tower (B).

Trees or other vegetation also cause turbulence. Again, a tall tower (C) is needed to raise your wind machine into smoother currents. Once you're away from the disturbance, you can use a shorter tower (D).

You'll need a very tall tower (E) to reach the best airflow if your wind machine is in a low spot.

Sizing Your Wind System

When you're sure that your site gets plenty of wind, it's time to decide how large a machine you'll need. As pointed out in Chapter 1, electric power should be used only where cheaper forms of energy can't do the job. Table 1-13 in Chapter 1 lists electricity use by various appliances, lighting, and other uses.

Check which appliances are heavy consumers of electricity, and which use only a little. Notice how quickly air conditioners, refrigerators, clothes dryers, and freezers run up the kilowatt-hours. A 5-kilowatt wind machine won't be able to handle peak loads for a modern all-electric home—even if the home isn't heated electrically.

If you're considering wind power for a small vacation cabin, you can keep power demand low and thus get by with a small wind machine at reasonably low cost. This means lights, radio, TV, and a few other low-power appliances. Maybe a small, efficient refrigerator, but no dishwasher, washing machine, or dryer.

If you plan to provide all the electricity for an average-size residence, you'll need 500 or more kilowatt-hours a month. This requires a 5-kilowatt wind machine with a 12- to 16-foot propeller.

Buying and Selling Electric Power: The Utility-Connected System

Most people get their electricity from a central utility miles away, and many people with wind machines still rely on the utility

F   G   H

Houses and other manmade obstructions can cause more turbulence than natural wind blocks, so nearby towers (F) must be located to avoid violent currents.

When the wind is blowing left to right, as shown here, a wind machine located at an upslope (G) can take advantage of fast-moving currents. But when the wind reverses, the tower will be on the lee side of the hill, engulfed in turbulence.

A tall tower on a hilltop (H) takes advantage of the extra altitude and places the wind machine into some of the fastest air currents of all.

for windless periods, or when they need more electricity than the wind can provide. Such a connection with a utility is a convenient arrangement; it eliminates not only the need for storage batteries, special wiring, and DC appliances (because batteries can produce only DC current), but the work that goes with them as well. The connection also makes it possible to not only buy electricity, but also to put wind-generated electric power into the utility grid when you have more than you need, and get paid for it.

During the energy crunch in the early 1970s, Congress passed the Public Utility Regulatory Policies Act (PURPA). The purpose of PURPA was to encourage utilities to supplement diminishing energy supplies with whatever other sources are available. Specifically, all utilities regulated by public utility commissions must buy electric power from a "qualifying small power production facility"—and sell power back to them at reasonable rates. Rural electric cooperatives may also come under this requirement, if the public utilities commission so decides.

"Qualifying small power production facility" means any system producing less than 30 megawatts of electricity. Wind machines produce only a tiny fraction of that amount, thus qualifying owners of wind machines as sellers of electric power for reasonable rates. "Reasonable" has been defined as the "avoided" cost of electric power, usually the most expensive power the utility is buying. Understandably, the power you generate must be suitable for the utility grid, and it must not create a safety hazard.

PURPA also states that a utility can't bill a small power producer the demand charge formerly levied on those who installed solar energy

smooth air

turbulent air

DETERMINING THE HEIGHT FOR A WIND TOWER

In order to get the most from your wind machine, it must be on a tower tall enough to reach the smooth steady air above gusty or erratic wind currents. A way to find the free-flowing air at your site is to fly a kite with a ribboned string. You can see the turbulence in the movements of the ribbons, and thus can determine how high your tower must be to reach the calm airflow above, where the ribbons blow straight.

equipment, a windmill, or other on-site power. As a result, many small power facilities can now afford to sell electric power to utilities. These include small conventional power plants, small hydropower plants, solar power plants, photovoltaic panels, and wind machines.

You won't need storage batteries if you connect to the utility, but you will need a synchronous inverter, a clever device that makes your wind power compatible with the utility line voltage and frequency. It automatically draws electricity from the utility grid when your house needs more than the wind can supply, and shuttles it to the grid when your wind system is generating more electricity than you can use.

Utility Regulations

Utilities are generally receptive to residential wind-machine connections. However, the utilities have rigid rules and regulations for the trading of electric power. You'll have to have power-conditioning equipment, special meters to measure how much you're buying and how much you're selling; you'll also need safety equipment. Safety features are critical so that utility repair people won't be endangered when working in the area. Most good wind machines incorporate proper safety features and should comply with utility regulations.

*There are two basic types of wind power systems: stand-alone and utility-connected. **Stand-alone** systems are self-contained. They use heavy-duty batteries to store excess power for later use. Battery banks should meet the peak kilowatt-hour demands of the home, and provide about five days' worth of storage. Some people play it safe and have a backup generator for emergencies.*

The batteries produce direct current (DC). DC can be used to run ranges and water heaters, but most appliances, clocks, stereos, and televisions run on alternating current (AC), so most stand-alone systems have an inverter to turn the DC into AC. If you have several appliances that can use DC, you can double-wire your home, with heavy-duty circuits carrying DC right from the batteries, and with separate wiring carrying AC from the inverter.

*Instead of supplying DC and AC, a **utility-connected** system has a synchronous inverter that matches your wind machine's output with the local line voltage and frequency. When the wind machine is producing exactly as much power as the home needs, no power flows between the home and the utility grid. When more power is used than the wind machine can supply, the synchronous inverter draws just enough power from the utility to balance your home's energy needs. And when the wind machine is producing more power than needed, the synchronous inverter feeds the excess power to the utility line for use by other customers.*

STAND-ALONE AND UTILITY-CONNECTED WIND SYSTEMS

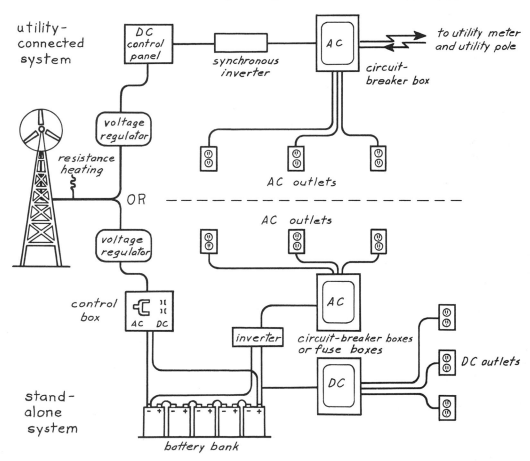

utility-
connected
system

DC
control
panel

synchronous
inverter

AC

to utility meter
and utility pole

circuit-
breaker box

resistance
heating

voltage
regulator

OR

AC outlets

voltage
regulator

AC outlets

control
box

AC   DC

inverter

AC

circuit-breaker boxes
or fuse boxes

DC

DC outlets

stand-
alone
system

battery bank

A stand-alone wind system may be the only power choice for out-of-the-way locations such as this, where running in utility lines isn't an option.

After safety, the utility will require that wind-generated electricity have the proper voltage and wave form so it won't cause any problems on the grid. Another regulation concerns the price to be paid for electricity—in each direction. Rates are generally set by the state public service commission.

Stand-Alone Systems

Stand-alone systems make the most sense—and sometimes are the only option—for remote locations where there are no power lines. These self-contained systems cost more and involve much more maintenance than utility-connected systems, but when you're faced with paying $10,000 a mile to run in power lines, they can be mighty attractive alternatives.

Stand-alones are more complicated systems because they need heavy-duty batteries to store power. And since batteries can produce only direct current (DC), you must have an inverter to change DC to alternating current (AC) or purchase special appliances that can run on DC.

For greater economy, you can double-wire your house, using heavy-duty circuits to carry DC from batteries for lighting and cooking, and separate wiring to supply AC power for the rest of your electrical needs.

Key to this kind of independent power is generally a 12-volt battery system. There are many advantages; for one, a 12-volt system is compatible with an automobile battery, so in an emergency you can get electric power for your house from the trusty automobile. Another advantage is safety. A 12-volt system isn't lethal, although, of course, you should be careful with any electric circuit.

Mobile homes pioneered the use of 12-volt DC power supplies; the recreation-vehicle (RV) catalogs are good sources of product information. If you plan a stand-alone system that must have 120 or 240 volts to run appliances vital to your life-style, your system must include an inverter to boost power from 12 volts to the higher voltages. Inverters cost money, and they use up some power, too. So if at all possible, get along on 12 volts DC. The first step in planning for stand-alone electric power is to determine how much electricity you really need. Remember: don't do anything with electric power that you can do with a cheaper energy source.

Storing Electric Power

The old water-pumping windmills still found on some farms and ranches generally can be counted on to pump a certain amount of water during a month or a season. But in order to have a continuous supply, the farmer or rancher must provide a water tank large enough to hold surplus water pumped during windy spells for use when there's little or no wind. The answer to a reliable supply of electricity is much the same, except that with a wind power system, electrical storage batteries are substituted for the water tank.

The more electric power used, and the longer the periods between good winds, the more battery storage capacity is needed. For example, if you use about 8 kwh a day and know that there may be two days with no wind, you must provide about 16 kwh of battery storage. Let's say that you have ten 12-volt storage batteries connected "in series" (this is explained a bit later) to give 120 volts for operating conventional residential appliances. The batteries you're using are rated at 150 ampere-hours. Multiplying volts times ampere-hours:

120 volts × 150 ampere-hours = 18,000 watt-hours, or 18 kwh

This battery bank provides a little more than two days worth of storage for windless days. But drawing all the energy from a battery constitutes a full cycle and shortens battery life. If you limit the drain to partial cycles, in which most of the stored energy remains in the battery, the battery will last up to three times as long. In the example

*Batteries connected in **parallel** (positive terminals connected to positive terminals and negative connected to negative) provide increased ampere-hour capacity while maintaining the same voltage.*

*Batteries connected in **series** (positive terminals of one battery connected to negative terminals of another battery) provide increased voltage while maintaining the same ampere-hour (storage) capacity.*

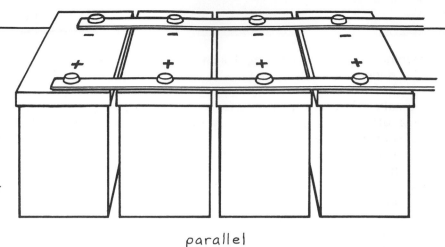

parallel

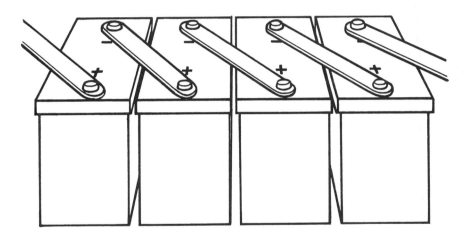

series

BATTERY BANKS AND CONNECTIONS

above, it would be economical in the long term to double the number of batteries to provide 36 instead of 18 kwh of storage.

Batteries are rated by the manufacturer to last a certain number of cycles, say 2,000. If drained below 50 percent capacity every day, they'll last 2,000 days, or about 5½ years. Using twice as many batteries—to provide twice the storage so that only partial cycling is needed—extends battery life to 12 or 15 years. This saves money over their lifetime and eliminates early replacement of the batteries. Of course, it also means a larger initial investment in batteries.

Multiply the price of a car battery by 10 or 20 for an idea of the minimum cost of providing storage battery backup for a small wind machine. A battery providing 1 kwh of storage costs from $75 to $150, depending on how good it is.

The battery bank drawing shows batteries connected in series and in parallel. Connecting batteries in series increases their voltage or power output. Parallel connection doesn't increase power but makes the batteries last longer.

Don't use automobile batteries to store electricity from a wind machine. The battery in your car is designed to provide a great amount of power for starting but won't last more than a year or two under the charge-discharge cycles of backing up a wind machine. Industrial deep-cycle batteries designed for longer life under frequent charge-discharge cycles can last 15 to 20 years. While first cost is higher, a good deep-cycle industrial battery far outlasts an automotive battery.

Maintaining enough storage batteries to back up a residential wind-electric system requires some work. The specific gravity of the acid, which indicates the charge of a battery, must be checked regularly with a hydrometer unless you have a voltmeter and electronic battery indicator gauges. There's little danger that batteries will freeze, so long as they are never allowed to discharge completely; battery acid acts as antifreeze. A fully charged storage battery is protected to about −90°F.

Maintenance

A wind machine is buffeted by gusty winds, baked by the sun, drenched by rain, and sometimes frozen by snow and ice. Components are subjected to high rotational speeds, vibration, and heating and cooling cycles. Obviously, such a machine must be properly maintained if it is to deliver electric power into your house for the 20 years or more you want it to last. The answer is regular and proper maintenance, either by you or on a service contract basis.

Wind machine maintenance consists of regular inspection and lubrication—generally every six months. Shafts and bearings must be oiled and checked for wear. Overspeed controls and generator components must be inspected for proper operation and wear, and electrical connections tested for integrity. Bolts must be tightened and blades examined for cracks or other indications of trouble.

System components on the ground also require frequent maintenance checks. Battery systems and wiring must be maintained and replaced periodically. If an inverter and utility power-conditioning and safety equipment are used, they must also be kept up.

Batteries will take the lion's share of your maintenance time. Check the battery charge often with a voltmeter; check the condition of the batteries monthly with a hydrometer and add water when necessary. (Some batteries are sealed and need no servicing during

Rotors must have a speed-control system to prevent them from overspeeding in high winds. There are many mechanisms to limit the rotor speed, including sideways and tilt-up governors, coning, and tip flaps.

With a **sideways governor**, the generator is set slightly off-center on its mounting, and the tail vane is hinged where it connects to the generator. In brisk winds, the rotor/generator starts turning away from the wind while the tail vane stays in place. A spring restrains the tail vane.

A **tilt-up governor** has its generator mounting hinged so that the rotor/generator tips upward in brisk winds. The restraining spring limits the tilting action until the wind pressure on the rotor overcomes the force of the spring.

In **coning** speed control, the blades bend away from the high-speed winds. This design is workable only on downwind machines; otherwise, the blades would hit the tower.

Tip flaps extend from the ends of the blades, tipping to create drag that slows down the rotor.

ROTOR SPEED-CONTROL MECHANISMS

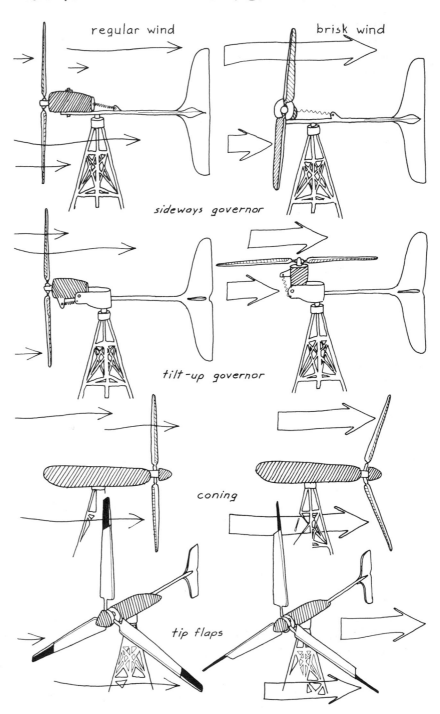

regular wind   brisk wind

sideways governor

tilt-up governor

coning

tip flaps

their lifetime.) Keep corrosion from building up on the battery terminals just as you do on your car battery. Once in a while, the batteries should be slightly overcharged to equalize the hydrometer readings of all the cells. A good wind will do this automatically. Donald Marier, author of the book *Wind Power for the Homeowner* (Rodale Press, 1981), suggests the following battery maintenance routine:

Weekly
>   Check battery bank voltage every one to three days

Monthly
>   Give batteries an equalizing charge
>   Check level of battery acid and hydrometer reading of a
>   few cells
>   Add water as needed (typically two or three times a year)

Yearly
>   Check hydrometer reading of all cells
>   Check all battery connections
>   Clean tops of batteries

Buying a Wind Machine

If you've run through all the pros and cons of installing a wind machine and feel that wind is for you, it's time to consider your purchase. Tackle the rebuilding of an old windmill, or the building of a new one from scratch, only if you're a competent mechanic with plenty of tools and the know-how to do the job properly.

In the 1960s and '70s, there were only a few new wind machines on the market, so many people chose to find and refurbish old windmills. Now that there are dozens of good commercial machines available in a range of sizes, types, and prices, refurbishing or building from scratch isn't as popular an option.

Typical listings of residential wind machines start with Aeolian and end with Zephyr. Power outputs range from ½ kilowatt to 25 kilowatts, with old-timers like Jacobs, Dunlite, and Sencenbaugh rubbing elbows with newcomers like Aero-Therm, Enertech, Milville Hawaii, and North Wind. Prices (1983) range from $545 for ½-kilowatt to $27,000 for 25-kilowatt systems.

Renewed interest in wind machines in the 1970s resulted in all sorts of unusual designs. A few have survived: one company markets a version of the Savonius rotor, one a Darrieus eggbeater type, one a gyromill, and two offer the bicycle wheel. But most commercial

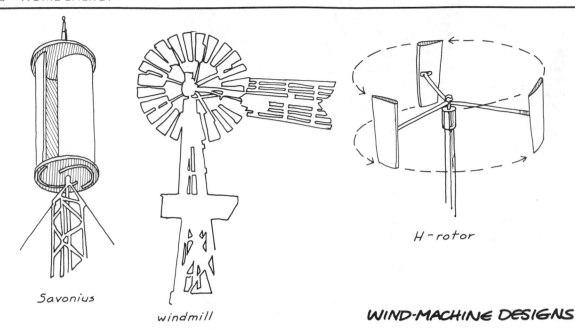

Savonius

windmill

H-rotor

WIND-MACHINE DESIGNS

If you live in an area with relatively low-speed winds, horizontal-axis wind machines are your best bet. These include the old-fashioned windmill, the H-rotor, and the propeller.

If, however, you live where the winds are gusty and frequently change direction, vertical-axis wind machines are usually the most efficient. These include the Darrieus, Savonius, and Darrieus/Savonius hybrid.

wind machines are the tried and true two- or three-bladed horizontal-axis types.

Voltages range from 12 volts DC to 480 volts AC. The Jacobs, granddaddy of them all, is priced at about $20,000 for a 10-kilowatt system, installation included. Enertech, one of the new generation of machines, goes for from $11,000 to $16,000 for an installed 4-kilowatt model. The new Excel costs $19,000 with tower and installation— $12,000 for the wind machine only; it delivers 10 kilowatts at rated wind speed.

Making Your Own Wind Machine

If you pass up the ready-made models, you still have a couple of options. The best are to rebuild an old windmill or build a wind machine from plans or a kit. Remember that the general planning process—siting, sizing, and backup system—is much the same whether you buy, build, or rebuild. Try to size the wind machine to your needs rather than take something just because it's available, appealing, or cheap. You may wind up cramping your life-style in trying to adjust to its output.

If you can locate an old Jacobs or Wincharger machine in fixable condition, if you know something about mechanics and electricity,

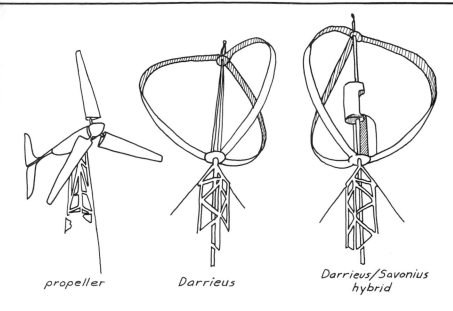

propeller   Darrieus   Darrieus/Savonius
                                            hybrid

and if you love to tinker, rebuilding may be the way for you to go. Start by getting a book or two on the subject—several are listed at the end of the chapter.

Plans or Kits

If an old machine is unavailable, or unappealing, you can buy a set of plans or a complete kit of parts and build your own wind machine. Such a project can be far more rewarding than merely writing a check and watching a professional crew put you into the electric power business. It may also be cheaper. Some plans are for wind machines so small they won't generate much power, although they may give you a good idea of what wind power's all about.

For More Information on Wind Power

Windpower: A Handbook on Wind, by Energy Conversion Systems, Daniel V. Hunt, 1981: Van Nostrand Reinhold, 135 W. 50th St., New York, NY 10020

For those who want the biggest, most complete book on wind. In addition to basic theory, installation, and applications, this one covers commercialization, the federal wind program, international efforts, and wind power's future.

The Wind Power Book, Jack Park, 1981: Van Nostrand Reinhold

> Not just the fundamentals, but how to build a wind machine for just about any application from water pumping to electric power to a wind furnace.

Wind Power for the Homeowner, Donald Marier, 1981: Rodale Press, 33 E. Minor St., Emmaus, PA 18049

> Written by an expert, it tells all about wind, wind machines, installation, storage batteries, connecting to the utility, appliances —everything but how to build a wind machine.

Suppliers

Prices and addresses given are as of this writing; check with the supplier before sending money.

Wind systems are not exactly do-it-yourself projects for most people, so there are few kits and plans to recommend. Had you checked the literature in the early '70s, you would have found many more, but time and experience have shown us that most of those early kits and plans were too difficult for owner-builders, or were just poor designs, doomed to failure. Here are one plan and one kit that survived the '70s.

Brace Research Institute (McGill University, Box 900, Ste. Anne de Bellevue, Quebec, Canada H9X 1C0) offers a plan titled "How to Construct a Cheap Wind Machine for Pumping Water." The plan is for a Savonius rotor, and it can be made from an oil drum. The plan costs $1.50 plus $2.50 postage.

Dragonfly Wind Electric (Box 57, Albion, CA 95410) offers the Dragonfly kit for an 8-foot windmill. Unassembled, unpainted, and unbalanced parts are sold as a complete machine, or can be bought separately. The kit costs $475 plus shipping.

Photovoltaics

Photovoltaic (PV) cells, popularly called solar cells, are thin wafers of specially processed silicon that generate an electric current from sunlight. One 4-inch-diameter solar cell produces about 1 watt of electricity when bright sun is shining straight down on it. A typical 1-foot-by-4-foot PV module delivers about 40 of these peak watts of power. (See page 235 for the explanation of peak watts, or peak power.)

Like the batteries that store wind power, in Chapter 6, PV cells connected in series produce more voltage. When they are connected in parallel, the current increases. PV arrays can be assembled using this series/parallel wiring method to deliver any amount of electric power desired—enough to power a wristwatch, a house, or a city.

Bell Laboratories invented photovoltaic cells and first used them to power telephone lines in Americus, Georgia. From rural telephone lines, PV cells quickly ventured into space. The first U.S. satellite, the Vanguard, was fitted with a few PV cells that powered a small radio for the several years the satellite remained in orbit. Millions of cells were soon producing space power for everything from Bell's small Telstar communication satellite to NASA's huge Spacelab, which used 10 kilowatts of PV electricity—22,000 miles from the nearest utility power line.

The early success of photovoltaics in space was possible only because Uncle Sam could afford the very expensive new power supply. How does $1,000 a watt—$60,000 to light a 60-watt bulb—sound? That was the cost of photovoltaics in the 1950s; today it's under $10 a watt. That's a lot better price, but still no bargain except in remote areas, where bringing in conventional electricity could be even more costly.

What You'll Learn in This Chapter

- How PV cells make electricity with no fuel except sunshine (page 233)

- While PV residential use is still in its infancy, there are experimental PV-powered homes in Massachusetts, New Mexico, and Arizona, to name just a few states (page 237)

- Except for remote sites, photovoltaics is still more costly than utility-supplied electricity (page 241)

- Stand-alone systems rely upon battery storage; utility interconnects sell excess electricity to the power company, instead of storing it, and buy electricity back when photovoltaics does not provide enough (page 245)

- How to mix and match PV power with a liquid-fueled generator to make a hybrid electric power system (page 247)

- Like a collector, a PV array usually gets mounted on the roof and should be oriented within 15 or 20 degrees east or west of true south (page 253)

- A PV system requires little maintenance because there are no moving parts, but batteries and power-conditioning equipment need some regular attention, and there are some safety measures to follow (page 253)

- How a $500 PV stand-alone system can provide enough electricity for lights, radio, TV, and small appliances in a summer cabin (page 255)

The cost of PV cells continues to drop, thanks to worldwide competition and technical improvements. The efficiency of PV cells has more than doubled, and this means half as many as in early PV days are now needed to do the job. Improved manufacturing processes are cutting costs even more. As efficiency goes up and costs come down, the price per watt of PV electricity could be $2.50 or less by the end of this decade, making PV cells competitive with just about any kind of electricity.

There are very good reasons for the tremendous interest in photovoltaics by manufacturers in the United States, Japan, France, Germany, and many other countries. The world obviously needs a new source of electrical power, and nuclear's future is fraught with serious safety concerns and poor economics. Photovoltaics may very well be the best

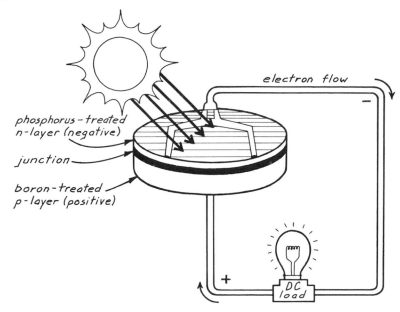

phosphorus-treated
n-layer (negative)

junction

boron-treated
p-layer (positive)

electron flow

−

+

DC
load

SOLAR CELL OPERATION

Photovoltaic (PV) cells convert sunlight into electricity. They are made of two thin wafers of silicon, one treated with boron, and the other treated with phosphorus, put together like a sandwich cookie. Phosphorus molecules have extra electrons, and boron molecules lack electrons, so an imbalance is created. High-energy electrons stimulated by light energy from the sun accumulate in the phosphorus-treated n-layer. Leads placed on either side of the cell enable the electrons to move from the n-layer to the boron-treated p-layer, completing an electric current.

alternative. Most PV cells are made of silicon, and silicon comes from sand, the most abundant element on earth. PV cells are clean and safe, producing no environmental hazards as they convert sunlight directly to electricity. They never require fuel, operate silently with no attention, and are very long-lasting.

How a PV Cell Works

You have to know something about physics to really understand how photons of light energy striking a PV cell cause the flow of electrons that make an electric current. Of course, you can still use PV electricity even if you don't understand how it works. But here's a brief explanation of how these seemingly simple little cells turn sunlight into electricity you can't tell from the kind that comes from the wall socket.

The basic mechanism of a PV cell is the silicon p-n, or positive-negative, junction. The p-layer is treated with boron to make it electrically positive. As a result, the boron molecules lack electrons (the

A typical 4-inch-diameter PV (solar) cell can produce about 1 watt of electricity.

tiny charged particles that make up the flow of electric current). The electrically negative n-layer molecules, treated with phosphorus, have surplus electrons, and this creates an "electrical potential" at the p-n junction.

When sunlight shines on the very thin n-layer, tiny bits of light energy called photons penetrate through it to the junction and dislodge electrons from some of the silicon atoms. This creates "holes" in the p-layer, and surplus electrons in the n-layer flow through external circuits to fill the holes. This electric current continues as long as light shines on the PV cell, and it can be used for a variety of purposes.

Except for electrons, there are no moving parts in PV cells, and thus nothing mechanical to wear out. Some PV cells have been operating for more than 20 years and are still producing electric power. And there's apparently no reason why they won't keep on doing that for years to come.

Most PV cells are round because single-crystal silicon is made by drawing a salami-shaped ingot from a furnace of molten silicon and then sawing it into thin wafers with special diamond saws. Many of these wafers, or cells, are interconnected to form a module. Typical modules produce 20 to 40 watts, enough to operate small appliances or to charge batteries. For larger amounts of power, modules are

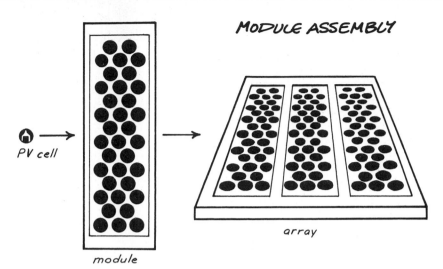

MODULE ASSEMBLY

PV cell

module

array

PV cells are connected in series or parallel, then sealed between protective glass or plastic to form modules. Modules are interconnected to form arrays that can be mounted on the ground or on rooftops to collect solar radiation. The direct current (DC) power produced by an array can be used by DC appliances or it can be stored. The DC can also be converted to alternating current (AC) by the use of an inverter.

interconnected to form panels, and then larger arrays. Arrays are mounted on the ground, on rooftops, or wherever they can best collect solar energy for a specific application.

Because silicon is fragile, PV modules are sandwiched between protective glass or plastic covers. Besides preventing breakage, this encapsulation also keeps moisture and contaminants from the PV cell. Some covers are so strong that the PV arrays beneath them can be walked on.

Square or rectangular PV cells are coming on the market, and new production methods will soon make actual cells the size of today's modules. In fact, one PV manufacturer produces thin-film cells that use microscopic amounts of material and are 16 inches wide and 1,000 feet long.

What Peak Power Means

You must understand the meaning of peak power to properly size a PV array to match electric power requirements. PV cells produce the same voltage regardless of their size or the amount of sunlight. But their power, or wattage, depends on cell size and sunlight.

When a solar cell is directly facing the bright sun on a clear day, it's delivering peak power. But when sunlight strikes a PV cell at an angle, the cell delivers less than peak power. (If you recall the solar basics of Chapter 2, you'll remember the same principle applies to flat plate collectors.) A fixed, properly oriented PV array produces peak power only at solar noon; early in the morning and late in the evening, the output is very low. As a RULE OF THUMB:

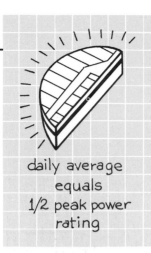

daily average
equals
1/2 peak power
rating

A PV cell produces a daily average of about half its peak power rating.

Another factor affects PV cell output. Peak power is actually achieved only when the air is very clear, when the sun is shining through a minimum of earth's atmosphere. A cloudy sky decreases PV power; when the sky is so cloudy or hazy that there are no shadows, a PV cell produces 25 percent or less of its peak power.

Nevertheless, PV modules can be counted on for useful amounts of electric power throughout the year. Here's a RULE OF THUMB:

A 1-kilowatt PV array generates from 1,600 to 2,200 kwh (kilowatt-hours) of electricity a year, depending on location.

1 kw array
generates
1,600-2,500 kwh/yr

And another RULE OF THUMB:

A house that doesn't use electricity for heating or air conditioning requires about 8,500 kwh a year as an average.

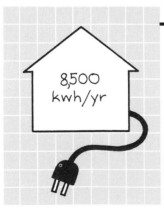

8,500
kwh/yr

From the above RULES OF THUMB, you can see that 4 to 5 kilowatts of peak PV power is needed to meet the electricity (but not heating or air conditioning) needs of a house. If heated with electricity, or air-conditioned with it in a place like Phoenix, a house uses from 10,000 to as much as 25,000 kwh a year and would need a proportionally larger PV array. Depending on insulation, sunlight, geography,

One of America's first PV residences, Solar One.

and other factors, including life-style, the PV array required for an average house ranges from 3 to 10 kilowatts of peak power.

State-of-the-Art Applications

One of America's pioneer PV residences was built in 1973 by Dr. Karl Böer, a solar scientist at the University of Delaware. Called Solar One, the house had a steeply slanted south-facing roof fitted with PV arrays to provide electric power. The arrays were designed to feed surplus power through a special meter to the utility during the day and use electric power from the utility at night. The PV panels also functioned as flat plate solar collectors, and fans blew warm air from the PV panels to phase-change heat-storage bins.

Solar One was a decade ahead of its time and did not achieve the goals of its builders. Nevertheless, it was a noble beginning.

As the price of PV cells dropped, functional utility-connected PV roofs began to appear in urban areas. One of the first was the John F. Long subdivision house in Phoenix, Arizona, funded mostly by the Department of Energy (DOE). The house's south-facing roof is covered with PV cells that generate about 7 kilowatts of peak power, enough to run a mechanical air conditioner plus appliances and lights—and still provide a monthly average of $45 worth of surplus electricity to the Salt River Project utility.

The John F. Long house in Phoenix has a utility-connected PV system.

Other PV houses quickly sprang up around the country, among them the Carlisle, Massachusetts, house designed by Steven Strong's Solar Design Associates (SDA). Its PV array, also funded by DOE, provided about the same power as the John Long house. Other SDA designs include the first privately funded PV residence, in Milton, Massachusetts; the privately financed El Dorado house in Santa Fe, New Mexico; and a 4½-kilowatt stand-alone (i.e., no utility backup) PV house in New York State's Hudson River valley.

In 1982 Volkswagen of Germany began testing PV arrays mounted on the roofs of Dasher station wagons. The panels didn't provide motive power but generated 160 watts of electricity for the ignition system.

Early in 1983, a specially built PV-powered car drove across Australia, covering more than 2,500 miles in less than 20 days. The 1-kilowatt PV array pushed the car to a maximum speed of 45 miles an hour, with an average for the trip of 15 miles an hour. Later in the year an American PV car drove from Long Beach, California, to Daytona Beach, Florida, a distance of 2,500 miles, in 18 days.

An even more exciting demonstration of PV electricity is the remarkable Solar Challenger, a full-sized, piloted airplane. The solar

The same PV cells that circle the earth on satellites have terrestrial applications as well, such as powering this experimental Volkswagen car. Other experiments include the Solar Trek car, which journeyed across Australia on PV power; and the PV-charged Solar Challenger plane, which was flown across the English Channel.

You don't need a warm climate to take advantage of photovoltaics—just plenty of sunlight. These homes are in Massachusetts.

plane has climbed as high as 14,000 feet on solar power, and it has made a remarkable 165-mile flight from France to England across the English Channel. Solar cells on its wings and horizontal tail powered electric motors that turned a large propeller.

Economics

The federal government, which funded much of the early PV research and demonstration programs, made overly optimistic projections of the cost of PV power: actual cost lags several years behind those early projections. However, the PV industry continues to insist that by 1990 PV electricity will be competitive with utility grid power on a strict dollars-and-cents basis, even without taking into account photovoltaics' many other merits. One or more of those merits may tip your own particular scales in photovoltaics' favor:

Cost-effectiveness in any size
Reliability
No fuel requirement
Stand-alone capability

Minimal maintenance
 requirements
Silent operation
No pollution

Space applications got the PV industry started; remote terrestrial applications have kept it going. At the 1983 price of $10 or less a watt, PV already competes with remote stand-alone systems that power, for example, summer cabins and communication stations. At $3 a watt, photovoltaics will replace gasoline or diesel generators anywhere. And at $1 a watt, photovoltaics will be so cost-effective that houses in this country will rapidly be roofed with silicon. Until then, it's tax credits that make many retrofit installations possible.

A giant leap for photovoltaics occurred in late 1982, when the ARCO Solar 1-megawatt utility power plant in California began feeding electricity into the Southern California Edison grid. ARCO soon afterward began a 16-megawatt plant for Pacific Gas & Electric (PG&E), also a California utility, and work continued on the first megawatt of the Sacramento Municipal Utility District 100-megawatt PV power plant.

When completed, the 100-megawatt plant will provide electricity for about 40,000 homes. ARCO delivered those cells for less than $5 a watt, the lowest price to date. Volume sales for utility power plants lower the costs of all PV cells and hasten the arrival of cost-effective residential photovoltaics—even without tax credits. In the meantime, the credits are nice to have; check Appendix C for details.

PV Power: How Powerful Is It?

Unlike the wind, the availability of sunshine is a fairly well known quantity. Solar radiation data have been collected for decades, and the averages that have been calculated do have significance for predicting the future. That's how solar water heaters are sized and solar homes designed. You can use the same data if you're thinking about solar cells for the family rooftop. There are at least three "How much?" questions involved: How much will they cost? How much will they produce? and How much is the PV output worth? The latter two numbers can vary somewhat depending on your location. The map reproduced here divides the country into a few general zones, each receiving a different amount of solar radiation annually. The numbers in each zone equal the approximate output of a PV system in kilowatt-hours per square foot per year (kwh/ft²/yr). There can, of course, be variations from place to place within a zone, but you can use these generalized numbers for a start. There can also be wide variations in the cost of the utility power that PV output is replacing.

You can compare the value of PV output with current or future costs for utility electric power by using the following equation:

$$\text{kwh/ft}^2/\text{yr} \times \text{\$/kwh utility} \times \text{ft}^2 = \text{\$/yr}$$
$$\text{(system output)} \quad \text{(cost of utility power)} \quad \text{(system size)}$$
$$\text{(annual value of system output)}$$

You can also find out how low the price for a PV system must be in order for it to give a favorable return on the investment. (For some people a five-year payback is acceptable, while others won't accept anything less than a three-year return. It's more of an individual choice.)

If, for example, utility power costs 10¢ per kilowatt-hour, a 100-square-foot PV system delivering 30 kwh per square foot per year would have to cost a maximum of $10 per square foot to pay for itself in 3⅓ years (100 ft² × $10/ft² = $1,000 system cost; 30 kwh/ft²/yr × $0.10/kwh × 100 ft² = $300/yr). The simple payback period is equal to the system cost divided by the annual return, or, in this case:

$$\$1,000 \div \$300 = 3.33 \text{ years}$$

Is Photovoltaics for You?

The time is coming when PV residential electricity will be standard equipment. That time, however, is still some years away, because, save for exceptional cases, such as remote sites, PV costs are con-

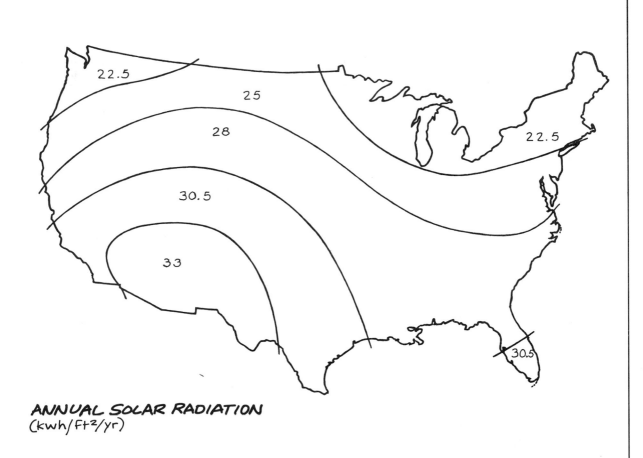

ANNUAL SOLAR RADIATION
(kwh/ft²/yr)

If a 5-year return on investment were acceptable, what would the maximum system cost be if all the above variables remained the same? You would simply multiply $300 times 5 years to find an allowable system cost of $1,500, or $15/ft² for the 100-ft² system. Of course, the current prices for PV systems are somewhat higher, though getting lower year by year. Meanwhile, the cost of utility power rises steadily.

The map shows average annual kilowatt-hours of sunlight falling on 1 square foot of ground or horizontal surface for different parts of the country. The figures are for a collector tilted at 45 degrees and facing south.

siderably higher than utility-supplied electric power. However, for some people the poor economics are overshadowed by the excitement of something new and the simplicity of extracting kilowatts directly from sunlight.

The Geography of PV

If you want to use photovoltaics for a summer residence, you'll be using PV electricity during the best solar part of the year. But if you plan to retrofit the house you live in year round, you'll have to consider seasonal variations in sunshine and electric power use. Look at the map of average annual sunshine in the United States for a preliminary estimate of how your location stacks up as a potential PV site.

There's more to PV geography than this, however. Heat lowers the efficiency of PV cells, a limitation that favors cool temperatures and works against hot desert regions. This doesn't mean that photovoltaics won't work in Arizona, of course, because it works there very well. But if another site receives as much sun and is cooler, PV performance will be that much better. For example, the air is cooler at higher altitudes, and higher altitudes generally are freer of pollution and allow more sunlight to reach PV cells.

The importance of solar access has been discussed at length in earlier chapters on solar heating. Solar access is even more important with PV power because of the higher cost of this solar technology. Furthermore, a shadow on even a portion of a PV panel can knock out much of its electric power production. So be very sure that you have access to the sun all year round, or at least when you plan to rely on PV power. Use the techniques described in Chapter 2, page 68, to check for possible shading.

How Much to Use?

A 15-kilowatt PV array is way beyond the means of most people. A 1-kilowatt installation is much more affordable—especially with a tax credit of 40 percent. Deducting $4,000 from $10,000 lowers the cost significantly; the same amount hardly makes a dent in $150,000.

Even with bright sun every day, a PV installation does not necessarily qualify as a cost-effective option. Compare the total cost (including system costs, tax credits, interest payments, and operation and maintenance costs) for 20 years of PV system operation with your present and projected electric power costs. A PV system installed on a house today could easily produce electricity as costly as 40 cents per kilowatt-hour. Since the average cost for utility-produced electricity is less than 10 cents per kilowatt-hour (the highest rate in the country is about 16 cents), a PV system in all likelihood will not be cost-effective even if any excess electricity produced is sold to a utility company.

However, as the price of PV systems comes down and the price of utility-produced electricity rises, future years should see cost-effective PV systems being installed on houses throughout the country. It is important to point out that if your house is in a remote, off-grid location and you use a generator to produce electricity, a PV system may be cost-effective right now on its own merit.

If you plan to rough it in a PV vacation home, you won't be concerned about backup electric power. But if you need supplemental power, be sure it's reliably available at a reasonable price. A utility connection, similar to that discussed for wind in Chapter 6, may work if you're on the grid and the utility is friendly. This lets you start with a small PV array and add more modules as you can afford them.

Refer to page 216 in Chapter 6 for a full explanation of utility connection regulations and the Public Utility Regulatory Policies Act (PURPA), which requires utilities to buy electricity from qualifying small independent power facilities such as home wind generators and PV systems.

Planning a System

In the decade since the experimental Solar One PV house, thousands of successful residential systems ranging from a few modules to 10-kilowatt arrays have been installed. Still expensive, such projects are nevertheless becoming more practical as suppliers offer off-the-shelf components for just about any size and type of PV power system.

Planning a PV installation is similar to planning other solar projects. Be sure you have adequate sunlight. Decide how much electric power you really need, using it only for those tasks that other energy sources can't handle at less cost. You wouldn't try to light your house with solar thermal collectors; don't try to heat your house or even your water with photovoltaics.

The Utility Connection

Why hook up to a utility if you're going to install PV panels? The wind machine owner does it because the wind doesn't blow all the time; the simplest answer here is that the sun doesn't shine all the time, either. Batteries store power but are expensive and require constant maintenance.

As with solar water and space heating, don't try for 100 percent PV solar power, because it's uneconomical to install a PV array and battery storage large enough to handle the worst possible weather. Utility backup allows a better energy life-style for a more reasonable cost than a PV stand-alone.

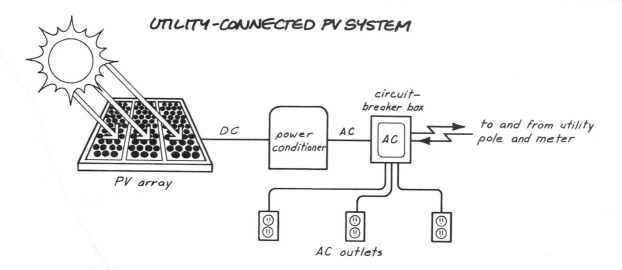

UTILITY-CONNECTED PV SYSTEM

PV array

circuit-breaker box

power conditioner

to and from utility pole and meter

AC outlets

In a utility-connected PV system any PV-generated energy that your house doesn't need is passed on to the utility. At night, on overcast days, or when you need more electricity than your PV system can give you, the system gets energy back from the utility. The power produced by the PV system is direct current (DC), which is usually converted to alternating current (AC) for use in the home. The power conditioner contains an inverter that converts DC to AC. The power conditioner also monitors the system's performance and regulates the power being passed on to the utility.

Some PV houses sell back more power to the utility than they buy. If PV arrays were cheap enough, we could all do this and make money. However, producing surplus power on a regular basis is evidence of an oversized PV array and unnecessary initial expense. PV power users stay with the utility for convenience, not to make money selling solar electricity.

In general, the same interconnection regulations apply for PV as for a wind machine. Electric power from a PV array must be properly conditioned to be sold to the utility. And safety equipment must be installed by a qualified electrician to protect you and others from any malfunction on the part of your power plant, or the utility. See page 208 in Chapter 6 for details.

Stand-Alone Systems

In spite of all that's been said above, there are far more PV stand-alones than utility-connected systems. People who can't hook up to utility lines are obvious candidates for stand-alone electric power, so there are thousands of small residential PV installations in rural or remote areas around the world.

There's another reason that's even more important to some PV adopters. As with a wind system, there can be a lot more user satisfaction in producing your own electricity from sunlight than in buying it from a utility. A PV stand-alone, like a wind power independent system, represents a return to the energy freedom of earlier generations —with the same freedom to do without when the sun doesn't shine or the wind doesn't blow!

The simplest remote PV applications use electric power when and as available to operate communication equipment, appliances, lights, and the like. More-useful systems store PV electricity in a small battery bank for later use as needed. Here's a battery-sizing RULE OF THUMB:

> For each watt of peak PV power, provide about 3 ampere-hours of battery storage.

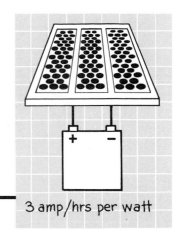

3 amp/hrs per watt

Thus, for a 35-watt array, about 105 ampere-hours of storage is recommended. This happens to be a standard battery size, costing from $75 to $150. (To convert to watt-hours, multiply ampere-hours by battery voltage. In this case, 12 volts × 105 ampere-hours = 1,260 watt-hours.)

In addition to batteries, two other components are needed with the electrical storage system. First, a blocking diode to let current flow from the PV array to the batteries, but not in the opposite direction. This keeps the batteries from losing power back to the PV array at night. Second, because sunlight varies in intensity, a voltage regulator or charge controller is put into the circuit to protect the batteries. ARCO Solar recently announced a PV module/storage battery system that requires no voltage regulator or charge controller. This improvement will reduce initial cost and simplify maintenance as well.

Hybrid Stand-Alone

This system is called a hybrid because in it a stand-alone setup is backed up with a generator that kicks on when sunshine is in short supply. The system has batteries, but fewer than a stand-alone, and there are fewer PV cells, too. It can get you started for a lot less money than a stand-alone can. To see why it's attractive, let's consider a stand-alone system with a daily electrical load of 5 kwh. It's designed for the worst conditions, since utility power just isn't available. This requires oversizing both the PV array and the battery bank at considerable cost.

With winter providing as little as 3.3 hours a day of peak power and a need for 10 days of battery storage, a PV array of 1.5 kilowatts peak power is required. This must be supplemented by a sizable battery

Table 7-1
Stand-Alone PV System

Components	Cost ($)
1.5-kilowatt PV array @ $10/watt	15,000
72-kwh battery bank @ $100/kwh	7,200
2,500-watt, 24-volt inverter	2,300
PV charge controller (if needed)	1,000
Total	25,500

array to provide the necessary backup. If the system is to use 110-volt appliances, an inverter is also required. This adds up to a very large initial expense (see table 7-1).

Deducting a $4,000 tax credit (the federal credit is $4,000, but if your state offers tax credits, the deduction will be even more) leaves a total cost of $21,500.

The stand-alone relies entirely on its photovoltaics and batteries for power. Laurence Jennings of Solar Electric Specialties, a California ARCO distributor, designed the photovoltaic gen-set hybrid system. It adds a liquid-fueled generator to provide electric power when needed, and reduces the size of the PV array and battery bank.

We can estimate the cost of this hybrid system to be $13,000 (see table 7-2). Again, deducting $4,000 in tax credits reduces the amount to only $9,000. Now you've got a hybrid system that provides power all the time for about one-third the cost of the all-photovoltaics system. An engine generator is noisy, and your PV system will still be dependent upon fossil fuel. However, you'll be using the fuel-powered backup generator for only about an hour a day. This compromise PV hybrid provides stand-alone power at a reasonable price. And as the cost of PV cells drops, you can add more modules and use the gen-set less and less.

Table 7-2
Photovoltaic Gen-Set Hybrid System

Components	Cost ($)
500-watt PV array @ $10/watt	5,000
4-kilowatt engine generator	2,400
Inverter	2,300
15-kwh battery bank @ $100/kwh	1,500
Battery charger	760
PV charge controller (if needed)	700
Manual transfer switch	225
Total	12,885

Worksheet 7-1
Appliance Power Consumption

Appliance	Watts	Hours/Day	Watt-Hours/Day
Blender	350		
B&W television	200		
Color television	350		
Fluorescent lights			
400 lumens	8		
865 lumens	16		
2,200 lumens	24		
Hot plate	600		
Incandescent lights			
865 lumens	40		
2,200 lumens	100		
Microwave oven	600		
Radio	40		
Refrigerator	200		
Small water pump	60		
Stereo	75		
Typewriter	30		
Washing machine	512		
Total			

Sizing the System

In general, the amount of money available determines the size of the PV system you'll buy. If money is not the issue, you can simply decide on the number of kilowatts you want and work backward from there. Most of us, however, have to do it the other way around, arranging electrical needs in order of priority and satisfying them off the top until the money runs out. Worksheet 7-1 lists common appliances and the watts of power they require.

Use the worksheet as an energy menu, selecting appliances and estimating how long they'll be used each day. Fill in the blanks, and add up amounts. The total watt-hours you come up with will tell you how large a PV array you'll need.

If your list comes to 500 watt-hours a day, how large an array do you need? First, let's get a handle on just how much PV modules

rafter

sheathing

direct

integral

ARRAY MOUNTING OPTIONS

can do. For finding the average daily power output of PV modules, use this RULE OF THUMB:

> Multiply peak output of PV array (in watts) by 5 to determine daily watt-hour total.

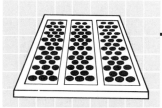

peak watt output x 5 =
daily watt/hrs

    The 5 represents average daily peak sunshine hours for the whole United States. The actual amount varies from about 3 hours in the Northeast to about 7 hours in most of Nevada. (Remember that peak hours represent only about half the total hours of sunshine.) See the map for peak sunshine hours in your region. On average, a 35-watt rooftop array in the central part of the country produces about 175 watt-hours a day, varying from about 105 watt-hours in the cloudier Northeast to 245 watt-hours in the sunny Southwest. Assuming you're in an average region, you'll get 175 watt-hours a day from each 35-watt module.

    Now you know how many watt-hours a day you want and how much a 35-watt module can provide. The next step is to divide your electricity requirements by the output of a 35-watt module: 500 ÷ 175 = 2.86. So three modules will give a trifle more power than you

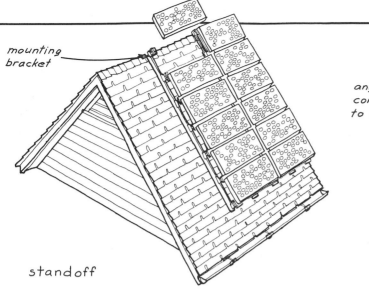

mounting bracket

standoff

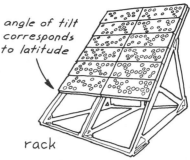

angle of tilt corresponds to latitude

rack

need. If the cost of three modules exceeds your budget, trim your wish list to the output of two modules, or 350 watt-hours a day.

Selecting Hardware

PV modules are manufactured in a variety of voltages and amperages, so it's easy to assemble panels to match your voltage and watt-hour requirements. The popular 35-watt module delivers a peak power of about 15 volts at 2.3 amps. Modules are connected in parallel for 12-volt stand-alone systems, or in series to produce higher voltage. Commercial PV modules tend toward standardization, so you can shop around for the best buys. However, try to deal with a firm that's knowledgeable and can be counted on for continuing service.

Mounting structures for PV arrays generally go on the roof. If the pitch of your roof (see page 74 in Chapter 2) is close to the tilt required for efficient collection of solar radiation, the array can be mounted parallel with the roof. Remember that the penalty for off-angle orientation isn't severe and that rooftop mounting keeps the PV array out of the reach of children, rocks thrown by lawnmowers, and other hazards.

Ground mounting allows exact southerly orientation and the opportunity to seasonally adjust array tilt for maximum performance. Modular mounting racks for rooftop or ground mounting of PV arrays are available.

*PV arrays should be mounted at the angle corresponding to the latitude of your location. You have several roof-mounting options: In a **direct mount,** PV modules are nailed right to the roof sheathing, replacing the shingles. There is no framing required.*

*If you're building a new home or an addition, an **integral mount** is a good choice because the PV array is mounted directly on the rafters, taking the place of the roof and eliminating the need for shingles, tar paper, and sheathing. Directly beneath the array is the attic space. You can provide array cooling by ventilating the attic.*

The stand-off mount *is placed several inches above the roof shingles, with mounting brackets penetrating the roof. Air is able to circulate behind the modules, providing back-surface cooling.*

*If you don't want to retrofit your roof, you can install a **rack mount.** Rack mounts are usually used for mounting arrays on the ground, though they are sometimes mounted on flat roofs.*

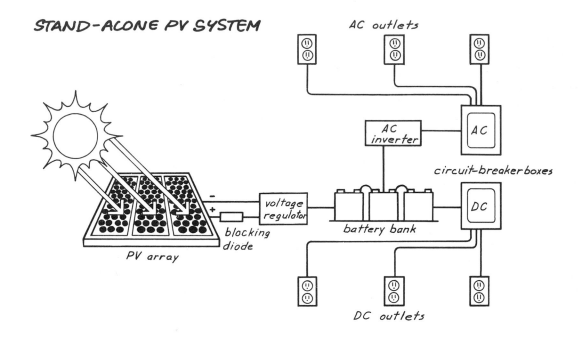

STAND-ALONE PV SYSTEM

AC outlets

AC inverter

AC

circuit-breaker boxes

DC

voltage regulator

blocking diode

battery bank

PV array

DC outlets

In this stand-alone PV system, a battery bank stores direct current (DC) power for later use. The DC power can be pulled directly from the batteries when needed, or, if you don't have DC appliances, the current can first be run through an inverter for conversion to alternating current (AC).

Battery storage is used to store electric power for nighttime and for those days when there's no sun. (People in the PV business insist there's no such thing as a no-sun day for photovoltaics because even on the darkest, coldest days an array collects some solar energy and produces electric power). Batteries aren't cheap, but the more of them you can afford, the less you'll be without electric power. Remember the PV manufacturer's recommendation of one 105-ampere battery for each 35-watt module. This gives 1½ to 2 days worth of storage without draining the battery excessively. If you can afford more batteries, use them.

Batteries specially designed for PV use are more expensive than ordinary lead-acid batteries. Antimony, calcium, or both may be added to the basic lead plates for better performance and lifetime. For example, Delco makes sealed batteries specially designed for PV applications that require no user attention. Exide makes PV-suitable tubular plate batteries alloyed with either antimony or calcium.

Power-conditioning equipment is necessary because a PV array doesn't provide the exact 12 volts DC or whatever AC voltage your system (or the utility) requires. We'll start with the simpler DC power systems (remember that recreational-vehicle suppliers are good sources for 12-volt equipment). Electric power from the PV array goes through a blocking diode and a voltage regulator, or charge controller, then to the battery bank, which in turn supplies electric power to the house.

A more sophisticated—and more expensive—AC power system is required if you use 110-volt lighting and appliances with your PV power system. Now the voltage regulator becomes a charge controller that protects the system better and makes it operate more efficiently. (Remember that some modules don't need charge controllers.)

Note than an inverter is added to the circuit between the DC storage battery bank and the AC load of the house. The inverter changes DC from the battery bank to AC for residential use. It also steps up the battery bank voltage to the required 120-volt or 240-volt AC line voltage. Depending on its quality, an inverter consumes from 10 to 30 percent of the power coming from the PV array. So don't skimp on the inverter; a good one will save money over its lifetime.

If you plan to sell back electric power to the utility, check with PV system suppliers—and the utility—in your area for the proper power-conditioning equipment.

Installing the System

Like a flat plate collector, the PV array should face as close to true south as possible. Although up to 20 degrees off-angle won't cause much loss in power, every improvement in electric-conversion efficiency increases the output of the expensive array. The advantage of correct orientation and tilt for a solar collector is discussed starting on page 63 in Chapter 2.

Seasonally adjusting a PV array to directly face the noon sun rather than leaving it at a fixed compromise tilt throughout the year adds about 4 percent to annual power output. In most cases, however, you'll have to settle on a fixed angle, generally equal to local latitude. With a roof mounting, also keep in mind such things as wind loading, snow packing, and accessibility for periodic cleaning and needed maintenance and repairs.

Maintenance

With no moving parts to wear out, a PV array requires little maintenance beyond an occasional washing to remove dust, and seasonal adjustment of tilt if you're using that option. A battery

A 35-watt PV system can be adequate to meet the limited electricity needs of a vacation home like this one.

bank will require regular maintenance; refer to page 223 in Chapter 6 for a refresher course. If the PV system is linked with a utility, the power-conditioning equipment and safety features must be kept in proper working order. An estimate of annual maintenance cost is from 1 to 2 percent of the system's first cost, with utility-connected systems cheaper to maintain than stand-alones.

Among the hazards for PV arrays are lightning and hail. Location and proper grounding procedures may help reduce the first danger. A protective cover would guard against hail for a fixed array. Maintenance of a PV residential system will also involve the infrequent replacement of voltage regulator, inverter, and batteries. So when buying these components try to be sure that replacements will be available in the future. The PV modules themselves should last 20 years or more unless defective, or damaged by hail or accident.

Table 7-3
35-Peak-Watt PV Module

Components	Cost ($)
1 35-watt PV module and mount	350
1 105-ampere-hour battery	100
1 blocking diode	1
1 voltage regulator (if needed)	50
Wire	1
Total	502

Three Ways to Go

With the fundamentals in mind, let's look at three typical residential PV applications of different sizes for an idea of the range of possibilities. First, a 35-watt stand-alone intended for a vacation cabin; next, a 500-watt stand-alone that will electrify a modest house; and then a 3-kilowatt utility-connected PV array.

35-Peak-Watt PV Module

Like the wind-powered stand-alone electric power systems discussed in Chapter 6, this small system uses 12-volt DC power that takes advantage of the variety of appliances and batteries available. Originally provided for RV and marine users, a variety of 12-volt appliances are now available from PV suppliers as well. These include high-efficiency refrigerators and other appliances, radios and TV sets, and fluorescent lighting, which uses only a fraction as much power as incandescent lighting. Table 7-3 gives the hardware you'll need for this system.

A 35-watt module is small enough to mount just about anywhere. Make sure it will be in the sun all day, all year. This little PV power plant provides a modest amount of electricity for radio, stereo, small black-and-white TV, lights, and small appliances. The power produced isn't enough to run all these at the same time, though; see worksheet 7-1 and table 1-13 to figure out how much electricity it'll take to power your appliances and lights and to work out a schedule of use. If the batteries are charged all week and used only on weekends, the PV system will do even more electric chores for you, limited only by the amount of battery storage. This is pretty good for a system costing only about $500 and requiring just about zero maintenance.

Table 7-4
500-Watt Array

Components	Cost ($)
14 35-watt modules (or equivalent) with mounts	5,000
14 105-ampere-hour batteries	1,400
1 blocking diode	1
1 voltage regulator (if needed)	500
Wire	5
Total	6,906

Table 7-5
3-Kilowatt Array

Components	Cost ($)
3 kilowatts of PV panels (plus mounts)	30,000
1 power conditioner	3,000
1 charge controller (if needed)	500
Total	33,500

500-Watt Array

This sizable PV array is almost 15 times as large as the little 35-watt starter set. Peak output power is 500 watts, a great deal more to work with. The array has an area of about 60 square feet and requires more planning and effort to mount and maintain. Table 7-4 gives the hardware needed.

Multiplying 500 peak watts by 5 peak sunshine hours gives about 2.5 kwh of electric power a day, 75 kwh a month. (Again, remember that the 5-hour sunshine figure is an average. Actual electrical output ranges from 50 to 100 kwh a month, depending on location and season.

This larger system adds considerably to the electric life-style possible: it could power a small, efficient refrigerator, additional appliances, maybe even a home computer once in a while, more and brighter lights, and color TV instead of black and white—not yet your typical all-electric house, but not primitive, either.

The 1983 cost for a 500-watt system totals $7,000 or less. With a system lifetime of 20 years, the cost will be about $500 a year, or $1.40 a day. If tax credits are available, the cost will be much less.

3-Kilowatt Array

The 3-kilowatt, 300-square-foot PV array provides an average of 10 to 20 kwh a day, 300 to 600 kwh a month. Not as much as the national average for electric power, but the owners of such a system can live almost like the folks down the road using utility power. And additional kilowatt-hours may be purchased when necessary or desired since the PV array is connected to the utility. Table 7-5 gives what you'll need.

At a cost of $30,000 for the PV array plus $3,500 for power-conditioning equipment, including an inverter to change DC to the proper AC voltage, the system isn't cheap. Over 20 years it comes to about $200 a month plus any supplemental electric power bought from the utility.

The Solar Electric Future

We're seeing today the beginning of one of the most rapid changes in energy use that has ever occurred. The PV cell is about 35 years old, its commercialization delayed because of very cheap conventional fuels. But two factors are strongly in its favor: the recent very rapid development of PV technology, and the increasing price of conventionally produced electric power.

Nuclear electric power was introduced just prior to the invention of PV cells; we were told then that the atom would bring us electricity so cheap we wouldn't have to meter it. That dream is dying rapidly, but there's a successful nuclear reactor a safe 93 million miles away that can beam power right to our rooftops.

For More Information on Photovoltaics

Photovoltaics: Sunlight to Electricity in One Step, Paul Maycock and Edward Stirewalt, 1981: Brick House Publishing Co., 34 Essex St., Andover, MA 01810

A good, basic overview of what photovoltaics is and what it can do. Also covers economic and societal aspects, and projects the future of PV applications.

Practical Photovoltaics: Electricity from Solar Cells, Richard J. Komp, 1981: Aatec Publications, Ann Arbor, MI 48106

A PV primer that will get you started toward designing your own system.

The Solar Electric Home: A Photovoltaics How-To Handbook, Joel Davidson and Richard Komp, 1982: Aatec Publications

All the necessary information for residential PV projects. Some science background will help you make best use of this book.

Suppliers

Prices and addresses given are as of this writing; check with the supplier before sending money.

PV Information and Equipment

Energy Sciences (Dept. 541, 832 Rockville Pike, Rockville, MD 20852) has an Energy Wonderbook catalog on photovoltaics and related equipment. This is a good introduction to the uses of PV cells, and excellent for hobbyists and experimenters. The catalog costs $3, refundable when you order an item.

Solar Electric Specialties Co. (Box 537, Willits, CA 95490) provides information on the gen-set and other useful PV applications. The company can also supply panels and other needs.

Solar Usage Now, Inc. (Box 306, Bascom, OH 44809) is a supplier of PV books, hobby projects, and equipment, including a PV-powered vehicle.

Western New England Solar (118 Maple St., Holyoke, MA 01040) is an excellent source of information and hardware for PV projects.

Appendix A
Heating and Cooling Degree-Day Maps

Cooling and heating degree-days represent the difference between the average outside temperature for a given day and a base indoor temperature of 65°F. (The base indoor temperature is a standard for calculating degree-days. It is not recommended that you keep your home as cool as 65°F in the summer.)

If the average outside temperature for a given day is 95°F, subtract 65° from 95° to get 30. This means that 30 cooling degree-days have accumulated for that one day. In cool weather you subtract the outside temperature from 65° to get heating degree-days. To determine the annual cooling or heating degree-days for your area, simply add up all the degree-days when the temperature outside was either above or below 65°F. The maps show the total annual cooling degree-days and heating degree-days for various parts of the country. The higher the number, the more cooling or heating you'll need to maintain a comfortable indoor temperature.

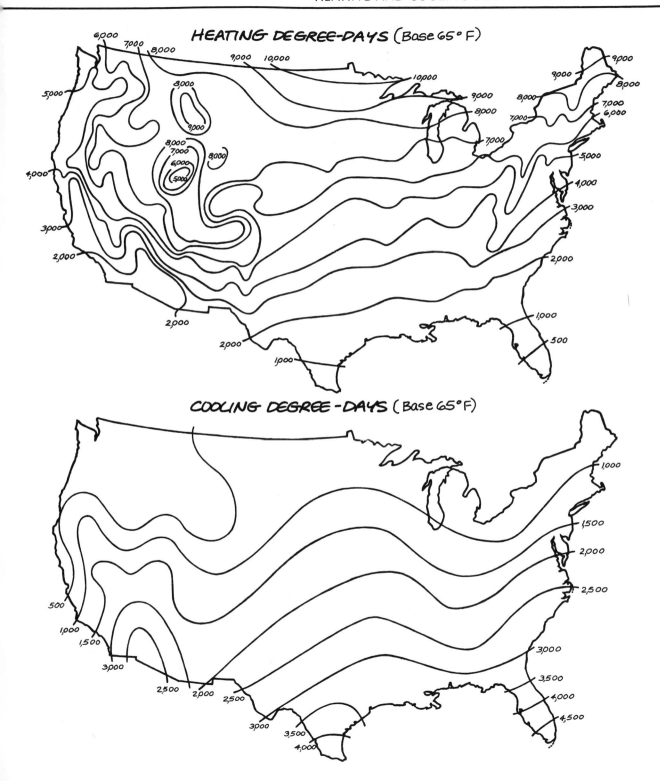

HEATING DEGREE-DAYS (Base 65°F)

COOLING DEGREE-DAYS (Base 65°F)

Appendix B
Computer Software and Applications around the House

After you've caulked and weather-stripped, solarized and retrofitted, is there anything left to say about making your house energy-efficient? Yes, and in a word—computers.

Personal computers aren't just for games and bookkeeping. There are programs that design energy-efficient houses and programs that design solar retrofits. There are even programs that will figure out how much money, if any, you'll save by retrofitting. Once you're in your home, your computer can monitor your home's energy performance. The easiest way to measure energy performance is simply to add up by hand the money you've paid out for fuel or electric power over a given length of time and balance that against the cost of the renewable-energy system you installed. However, this basic monitoring may not be as good as that which you can do with a computer because it may not turn up problem areas that could be corrected for better performance.

The key to performance monitoring is the installation of sensors to measure the amount of energy used, inside and outside temperatures,

humidity, wind velocity, solar radiation received, and other factors needed to establish the efficiency of your energy system. Some energy-monitoring systems include dozens of these sensors, all to be connected to your computer—which then records the information, sorts it out, and gives you an accurate account of how well your energy system is functioning, and just where your energy dollars are going.

With sensors in place, many energy-monitoring programs also allow you to regulate your furnace, keep a watchful eye on your water heater, and control your solar equipment. Other systems allow you to program your computer to turn lights on and off, open and close doors and windows, sense intruders, signal alarms, and automatically call the fire and police departments. And, with a voice synthesizer, a computer can even tell the kids a bedtime story.

Architects and engineers have been using computer-aided design for quite some time when designing industrial buildings. Until recently, however, there has been very little software available for residential applications. And until very recently, there have been no programs around for computerizing homes. The few homeowners who've taken the plunge have been hobbyists who have written their own programs and have installed their own sensors, fans, and other auxiliary equipment.

But all that is changing. As more people purchase personal computers, and as more and more people become interested in improving their homes' energy efficiency, their demands for energy-related residential software are being met.

What follows is a look at some of the home energy programs currently available.

Programs for Hand-Held Calculators

Princeton Energy Group (729 Alexander Rd., Princeton, NJ 08540) offers PEGFIX, which calculates auxiliary heat requirements and excess heat available in solar energy systems; and PEG-FLOAT, which calculates hourly temperatures of air and storage mass for glazing at any orientation. Each program costs $75.

Solarcon (607 Church Rd., Ann Arbor, MI 48104) offers a book of 13 solar energy programs for $50. The programs include ECON, a life-cycle cost estimate for conventional and renewable energy systems; F-CHART, a design of liquid or air space heating and domestic water heating; STORE, a program for storing and retrieving weather data; and DG, a design method for direct-gain passive systems.

Total Environmental Action, Inc. (Church Hill, Harrisville, NH 03450) offers TEANET III, a passive simulation program that calculates auxiliary energy requirements and hourly temperatures for up to seven points. It costs $95.

Programs for Personal Computers

Aeolian Kinetics (Box 100, Providence, RI 02901) offers information on computer hardware and software used in "Class B" residential energy monitoring.

Compu-Home Systems, Inc. (3333 E. Florida Ave., Denver, CO 80210) offers TOMORROW-HOUSE, a complete home control system, with hardware and software, for $895.

F-Chart Software (Beckman Duffie Assoc., 4406 Fox Bluff Rd., Middleton, WI 53562) offers F-CHART, which performs analysis and design of active and passive heating systems, for $400; and F-LOAD, which calculates monthly heating loads of buildings and life-cycle heating costs, for $425.

Howard W. Sams & Company, Inc. (4300 W. 62nd St., Box 7092, Indianapolis, IN 46206) offers COMPUTER-ASSISTED HOME ENERGY MANAGEMENT, a program that helps you create a working energy monitor for your home. The system can measure almost any form of energy consumption. It costs $15.95.

Londe-Parker-Michels, Inc. (7438 Forsyth, Suite 202, St. Louis, MO 63105) offers INSULATE, which calculates optimum insulation R-values, total investment, and energy savings; and OVER-HANG, which calculates shading effects of overhangs. The programs cost $195 each. PASODE calculates auxiliary heating requirements for water walls, Trombe walls, and direct-gain systems. It costs $295.

Princeton Energy Group (729 Alexander Rd., Princeton, NJ 08540) offers IMPSLR, which calculates solar savings fraction, internal heat gains, and required auxiliary heat, for $250; MICROFIX, which calculates performance of a direct-gain or attached sunspace passive system, for $150; and NEATWORK, which performs a two-zone simulation of direct-gain mass elements, water walls, and shaded mass, for $400.

Solarsoft, Inc. (Box 124, Snowmass, CO 81654) offers five programs for $995, or priced individually, as shown. The programs are F-CHART, which predicts solar contributions for active liquid and air collectors for space heating, domestic hot water, and pool heating, for $395; SOL-GAIN, which calculates clear-sky incident and transmitted radiation through single or double glazing at any orientation, tilt, or latitude, and specifies horizontal overhangs for shading, for $400 in a package with T-SWING, which calculates temperature swings in a building; SUNOP, which optimizes energy conservation levels and gives optimum solar saving fraction and passive collector area, for $250; and SUNPAS, which calculates solar energy for direct-gain, Trombe wall, water wall, and sunspace, for $250.

Solartek (RD 1, Box 255A, W. Hurley, NY 12491) offers SUNGRAPH, which calculates the sun's position in the sky for any time, for $49; SUNHEAT 1, which designs and evaluates solar water-heating systems, for $29; SUNSIM-2, which simulates flat plate tracking solar collectors, for $49; SUNSIM-3, which simulates concentrating tracking collectors, for $49; and SUNSIM-4, which calculates energy available from the sun for use in space heating, cooling, and water heating, for $59.

X-10 Interfaces

BSR (USA) Ltd. (Rte. 303, Blauvelt, NY 10913) manufactures the BSR System X-10, a remote-control system for lighting and appliances. Lights and appliances are plugged into lamp, appliance, and wall receptacle modules that are commanded from a compact keyboard. You can make the system more flexible by interfacing it with your personal computer. Then you can monitor your appliances' energy usage and tie in your lighting system with a burglar alarm system. The following companies make products that interface various computers with the BSR System X-10.

Connecticut Microcomputers (36 Del Mar Dr., Brookfield, CT 06804)

Interface Technology (Box 383, Des Plaines, IL 92103)

Radio Shack (1 Tandy Center, Fort Worth, TX 76102)

Sci-Tronics (523 Clewell St., Box 5344, Bethlehem, PA 18015)

Thunderware (Box 73322, Oakland, CA 94601)

Appendix C
Energy Tax Credits

There are two basic tax credits for users of renewable energy: federal credits and state credits. We'll cover the federal credit first, as it is the simplest and applies to everyone.

Federal Tax Credits

A qualified taxpayer is allowed to deduct 40 percent of the cost of solar, wind, photovoltaic, or geothermal energy equipment purchased for year-round home use since December 31, 1979. The maximum credit allowable is 40 percent of $10,000, or $4,000. Cost can include not only the renewable-energy equipment but installation costs as well. There are very strict requirements for passive solar credits, so check these carefully.

This $4,000 is a true tax credit, subtracted from the bottom line of your federal income tax form. If you don't owe that much tax in a year, the remaining credit can be carried over until you've recovered the full amount of tax credit. The homeowner files for these credits on Internal Revenue Service (IRS) Form 5695, as explained in IRS Publication 903, "Energy Credits for Individuals."

In addition, the federal government allows a business tax credit for installation of renewable-energy equipment. This is for 15 percent of the cost, up to $10,000, or a maximum of $1,500. While this is not as attractive as the 40 percent residential credit, it is certainly worth applying for. If you are in this category, check with your accountant, or contact the local IRS office for details. California also offers a business tax credit of 15 percent.

State Tax Credits

State credits are not as simple as federal credits. Most states offer them, but the credits differ greatly. Check with your state energy office for copies of tax credit legislation.

Before anyone yells "foul" at the idea of spending tax dollars on renewable-energy credits, it's fair to point out that the oil, coal, natural gas, and nuclear industries have been subsidized to the total tune of more than $200 *billion* over the years. These subsidies continue at the rate of many more billions each year, and as a taxpayer you pay for them.

That's the good news. The bad news is that these renewable-energy tax credits may be phased out over a period of years, with 1985 generally the last year they will be given—which is all the more reason for installing a solar water heater, and maybe some other renewable-energy hardware, while tax credits are still available. Check with your state energy office for details on state credits, and with the IRS for a copy of Publication 903 describing federal credits.

Glossary

absorber—the surface in a solar collector that absorbs solar radiation and converts it to heat energy; generally a matte black metallic surface

active solar energy system—an energy system that requires energy from an outside source to collect and distribute and/or store solar energy, usually by means of fans or pumps

airtight wood stove—a well-sealed stove that is more efficient than a "leaky" stove because the draft can be better controlled and the S-shaped baffling means less heat escapes up the flue; sometimes called a Scandinavian stove

alternating current (AC)—electric current that changes its direction of flow at regular intervals, normally making 60 cycles per second. Alternating current is easier to transmit than direct current and is also more easily changed to higher or lower voltages. Household current is AC

ampere—the unit of rate of flow in an electric current

ampere-hour—a unit of electrical charge, equaling the quantity of electricity flowing in 1 hour past any point of a circuit carrying a current of 1 ampere. Storage batteries are rated in ampere-hours to show the quantity of electricity that can be used without discharging the battery beyond safe limits

barometric damper—a weighted flue damper that, if adjusted properly, automatically provides the right amount of air for good combustion and minimal pollution

batch water heater—a water tank painted black, usually placed in an insulated box with a glazed cover; designed to both heat and store this heated water for later use

battery bank—a number of storage batteries connected together, providing the desired voltage in storage capacity

berm—a man-made mound or small hill of earth, as in bermed (or partially buried by mounds of earth) houses

blocking diode—a device that allows an electric current to flow in one direction but not in the opposite direction

Btu (British thermal unit)—a unit used to measure the quantity of heat; more specifically, the heat required to raise the temperature of 1 pound of water 1°F, approximately the heat given off by one burning kitchen match

catalytic combuster—much like a car's catalytic converter, a device that chemically lowers the temperature at which emission gases burn, causing the gas to combust. A wood stove with such a device (a catalytic stove) releases extra heat into the house while consuming more pollutants than the same stove without one

chimney effect—the characteristic of air to rise when heated because it has a lower density (and is therefore lighter) than the surrounding air; this characteristic is used to help cool a building by allowing warm air to rise up and out upper-story windows, attic vents, and thermal chimneys

circulating wood stove (or fireplace)—a double-walled stove (or fireplace) that, often with the assistance of a fan, circulates air between the walls and then throws this heated air out into the room

clerestory—a vertical wall that contains a window that is located there not so much to provide a view to the outside, but rather to provide light into a building

coefficient of performance (COP)—a measure that represents the number of Btus of heating or cooling provided for every Btu of energy used by an appliance

collector—see solar collector

collector angle or tilt—the angle at which a collector is tilted with respect to a horizontal plane, designed to maximize the collection of solar radiation

conduction—the direct transfer of heat energy through a material; for example, warm inside air heating cooler walls and window glass, which in turn transfer their warmth to the still cooler exterior surfaces and then to the outside air

convection—the transfer of heat by movement of a fluid, usually air or water; for example, warm inside air rising to the ceiling or upstairs

convective loop—see thermocirculation

cool tube—an underground pipe that channels naturally cool air from below ground level to the living area

cooling degree-day—see degree-day

cord of wood—see face cord, long cord, standard cord

creosote—an oily, odorous distillate of wood tar that may collect on the walls of a stove or flue as a result of incomplete combustion. Since creosote can ignite and cause chimney fires, stove flues should be cleaned regularly

damper—a flap that controls the passage of air through a duct or flue, such as in a woodstove flue and in a hot air duct

degree-day—the difference between the average outside temperature for a given day and a base indoor temperature of 65°F. A *heating* degree-day is one day with the average outside temperature 1° *cooler* than 65°F. A *cooling* degree-day is one day with the average outside temperature 1° *warmer* than 65°F. The total number of heating (or cooling) degree-days over the heating (or cooling) season indicates the relative severity of the winter (or summer) in that area

desiccant cooling—a method of cooling that uses trays of material such as activated charcoal to absorb moisture from the air

direct current (DC)—electric current that flows in one direction. Generators produce DC and batteries store DC, but most appliances use AC

direct solar gain—heating of a space by solar energy directly entering that space; for example, a living space warmed by sun shining through its south-facing windows

drain-back system—a type of freeze protection for a solar water-heating system that circulates antifreeze solution through the collector and into a heat exchanger in the water tank, where heat is transferred from the antifreeze to the water

drain-down system—a type of freeze protection for a solar water-heating system that is designed with controls that drain the collector of water when subfreezing temperatures occur

earth berm—*see* berm

energy audit—an accounting of the forms and amounts of energy used during a designated period; for example, an annual house energy audit

energy efficiency ratio (EER)—the amount of useful heating or cooling provided (measured in Btus) divided by the electrical energy input (measured in kilowatt-hours). EER ratings are found on some appliances and are useful in comparing relative efficiencies: the higher the EER rating, the more energy-efficient the appliance

eutectic salts—*see* phase-change material

evaporative cooling—a method of cooling the air by moving it over a wetted surface. In an evaporative cooler unit, a blower pulls outside air through a wet filter pad. The evaporation of water from the pad then cools the room air

face cord—a measurement of stacked firewood that is 4 feet high, 8 feet wide, and 1 to 2 feet long

fireplace insert—a mechanism that increases the efficiency of a fireplace by circulating cool room air around the firebox, where it is heated, and then moving it back into the room

flat plate collector—*see* solar collector

flue—a metal or ceramic pipe that vents combustion gases from a stove, furnace, boiler, fireplace, or wood heater to the outdoors; also known as a chimney, a stack, or a stovepipe

freestanding fireplace—a fireplace "in-the-round"; one that is open or glazed 360 degrees around and that sits anywhere in a room, away from a wall

generator—that part of a wind machine that converts the rotary motion of the propeller or rotor into electricity

glazing—glass or plastic windows, doors, clerestories, skylights, sunspaces, and collector coverings, used for admitting light and solar heat into a space

greenhouse effect—the trapping of heat behind plastic or glass glazing with a consequential rise in temperature due to many causes, such as reduced convection and reflection and increased absorption of heat. The term is used to describe the internal temperature rise of water or air behind solar collectors, or within a sunspace or greenhouse

heat exchanger—a device that transfers heat from one medium to another, such as from water to air or water to water

heat mirror—clear plastic film placed on or between panes of glass to increase the glass's insulating ability

heat pump—a device that removes low-temperature heat from a source (such as the atmosphere or water), concentrates the heat, then delivers the resultant high-temperature heat to a living area, to storage, or to a hot-water supply

heating degree-day—*see* degree-day

heating load—the amount of heat required to maintain indoor comfort, measured in Btus per hour

hot air box—*see* thermosiphoning air panel

hybrid solar system—an energy system incorporating both active and passive solar components; for example, a thermal storage masonry floor heated by hot water pumped from a solar collector

indirect solar gain—heating of a space by a material or by another space that has been heated directly by the sun; for example, a living space heated by a Trombe wall or by an adjacent sunspace

infiltration—the uncontrolled movement of outdoor air into a building through leaks, cracks, windows, doors, and other openings in the building

insolation—the amount of solar radiation striking a surface exposed to the sky, measured in Btus per square foot per hour (and sometimes in watts per square meter per hour)

inverter—a device that converts direct current (DC) to alternating current (AC)

kilowatt—a measure of power or heat flow rate, equal to 1,000 watts or 3,413 Btu per hour

kilowatt-hour—the amount of energy equivalent to 1 kilowatt of power being used for 1 hour

long cord—a measurement of stacked firewood that is 4 feet high, 8 feet wide, and more than 4 feet long

magnetic, or compass, south—south as indicated on a compass for any given location; it changes with longitude and is not usually the same as solar south

masonry wood-burning stove—see Russian stove or fireplace

megawatt—1 million watts

night, or movable, insulation—insulation that can be moved easily over glazing to reduce heat loss at night and during cloudy periods and that may also be used to reduce heat gain in summer; for example, insulated window panels and shutters, insulated drapes and shades

passive solar energy system—an energy system that uses the structure itself, rather than outside energy, to collect, store, and distribute solar energy

peak power—the maximum voltage. More specifically, in a photovoltaic system, the maximum power a PV cell can generate

phase-change material—a material such as eutectic salts or Freon that stores heat when it melts and releases heat when it solidifies. Because of their concentrated heat-storage capacity, phase-change materials are an alternative (albeit an expensive one) to masonry and water for thermal storage

photovoltaic array—a group of interconnected photovoltaic modules that can be mounted on the ground or on rooftops

photovoltaic cell—a device without any moving parts that converts light (such as sunlight) directly into electricity by the excitement of electrons

photovoltaic module—a group of photovoltaic cells connected in series or parallel and sealed between protective glass or plastic. Several interconnected modules make an array

R-value—a measure of the ability of a material to resist the flow of heat through it; the greater the R-value, the greater the material's insulating ability

radiation—the transfer of heat by electromagnetic waves, such as light and heat; for example, a hot wood stove sending waves of heat into the air around it

reflective film—mirrorlike plastic film placed over a pane of glass to either reflect escaping heat back inside (during cold weather) or reflect outside heat away from the glass (during warm weather). The direction of the reflection depends upon the placement of the film

retrofit—to fit solar heating systems to existing buildings; or more generally, any addition of a new technology to an existing structure

rotor—the propeller of a wind machine, moved by the wind, that turns the generator, which in turn produces electricity

Russian stove or fireplace—a massive masonry (usually brick) wood stove characterized by its long, serpentine baffling that channels hot air and gases over lots of thermal mass, which in turn stores the heat and releases it slowly to the room. Also called a masonry woodburning stove

Scandinavian wood stove—see airtight wood stove

selective surface—a surface that is a good absorber but poor emitter of solar heat; used as a coating for absorbers to increase collector efficiency

skylight—a glass or plastic panel installed in a roof to admit natural light, and sometimes solar heat, to a living space

solar attic—a solar space-heating system designed to heat inside attic air by means of south-facing roof glazing and then duct this air by fans to living spaces below

solar cell—see photovoltaic cell

solar collector—a device that collects and converts solar radiation into heat; for example, a flat plate collector, which is a glazed box through which water or air is circulated and heated

solar gain—that part of a building's heating load (and sometimes its cooling load) that is provided by solar energy

solar noon—that moment of the day that divides the daylight hours for that day exactly in half. To determine solar noon, calculate the length of the day from the time of sunset and sunrise and divide by two

solar rights or access—the ability to receive direct sunlight that is passed over land located to the south; the protection of solar access is a legal issue

solar south—see true south

stand-alone system—a self-contained energy system (such as wind or photovoltaic) that uses batteries for storage and is not connected to a utility system

standard cord—a measurement of stacked firewood that is 4 feet high, 8 feet wide, and 4 feet long

standby loss—the heat loss from a boiler or furnace through conduction or radiation; the better insulated the furnace or boiler, the less the standby loss

sun-tempered building—a building that collects a significant amount of solar heat but generally lacks a means of storing that heat (in thermal mass, phase-change materials, and the like)

sunspace—a living space enclosed by glazing; a solarium or greenhouse

synchronous inverter—a device in a utility-connected system that converts direct current (DC) from a wind machine or from photovoltaic cells into alternating current (AC). Getting its frequency signal from the utility grid, it draws electricity from the home system that is synchronous (compatible) with utility electricity

thermal mass—a wall, floor, or column of masonry (brick, concrete, tile, or stone) or water-filled containers designed to store heat from the sun and release it to living spaces. A Trombe wall and a water wall are specific types of thermal mass

thermocirculation—the convective circulation of a fluid, such as water or air, that occurs when warm fluid rises and is displaced by denser, cooler fluid in the same system; sometimes referred to as a convective loop

thermography—an optical measuring technique that produces color photographs showing where a house is losing heat

thermosiphoning air panel (TAP)—a passive solar space heater that heats air in a collector located below the living area so that the warmed air flows up to the living area by natural convection, without the assistance of a fan, and then back down to the collector, forming a convective loop. Sometimes called a hot air box or a window box heater

thermosiphoning water heater—a system in which water heated by a solar collector flows up to a storage tank by natural convection, rather than with the assistance of a pump, and then back down to the collector, forming a convective loop

Trombe wall—a thermal storage wall named after its inventor, Felix Trombe, that incorporates glass or plastic on the outside of a darkened wall. Warm air rising between the two surfaces can be ducted to the living space adjacent to it, or it can be stored in the wall for slow release to the living space

true south—the position of the sun at exactly solar noon or any given day at any given location; not necessarily the same as magnetic south

utility-connected system—an energy system (such as wind or photovoltaic) that is connected to the utility grid, enabling the system to draw (buy) power from the utility company and, if regulations allow, send (sell) excess electricity back to the utility company

utility grid—the electric power plants and transmission lines owned by a utility company

vapor barrier—a material that is impervious to the flow of moisture and air; if placed on the warm (usually indoor) side of insulating material, it prevents condensation in walls and other places where insulation is found; for example, the foil backing on fiberglass batt insulation, plastic film on the interior side of loose-fill insulation, and vapor barrier paint over plasterboard

volt—the unit of pressure in an electric circuit

voltage regulator—a device that regulates the amount of pressure (voltage) in an electric circuit

water wall—water-filled containers, usually plastic cylinders, tubes, or bottles, or rustproofed metal cans or drums, that are designed to absorb and store solar heat; a type of thermal mass

watt—the unit of rate at which work is done in an electrical circuit, equal to the rate of flow (amperes) multiplied by the pressure of that flow (volts)

window box heater—see thermosiphoning air panel

Photo Credits

Index

Page references in italic indicate photos, illustrations, and captions; references in boldface indicate tables.